Encouraging Girls in Mathematics

Encouraging

ΣπΣπΣπΣ Girls in ΣπΣπΣπΣ

Mathematics

The Problem and the Solution

LORELEI R. BRUSH

with the assistance of
Cynthia Char
Nancy Irwin
Glen Takata

Abt Books
Cambridge, Massachusetts

Q A13
B78

Library of Congress Cataloging in Publication Data

Brush, Lorelei R   1946–
  Encouraging girls in mathematics.

  Bibliography: p.
  Includes index.
  1. Mathematics--Study and teaching (Secondary)--
United States--Psychological aspects.   2.   Adolescent
girls--Education--United States.   3.   Mathematics--Study
and teaching (Secondary)--New England--Longitudinal
studies.   I.   Title.
  QA13.B78   510'.7'12   79-55774
  ISBN 0-89011-542-7

© Abt Associates Inc., 1980

Printed in the United States of America.

To Bill and Dick,
who encouraged me

Contents

LIST OF FIGURES   ix

LIST OF TABLES   xi

PREFACE   xiii

1 THE PROBLEM   1

    PROLOGUE   1

    DESCRIPTION OF THE PROBLEM   2

    EXPLANATIONS OF AVOIDANCE   9

    Lack of Ability   9

    Negative Attitudes Toward Mathematics   12

    Lack of Usefulness of Mathematics for Future Life   15

    A Discouraging Social Milieu   17

    SUMMARY   19

2 THE RESEARCH PROJECT   21

    THE STUDENTS   24

    THE INSTRUMENTS   28

    The Questionnaire   29

    Essays   35

    Interviews   35

    Standardized Ability Tests   36

    SUMMARY   37

3 STUDENTS' ATTITUDES TOWARD MATHEMATICS   38

    THE CLUSTER OF PLEASURE/DISPLEASURE WITH MATHEMATICS   42

    THE PERCEIVED USEFULNESS OF MATHEMATICS   54

    INFLUENCES FROM THE SOCIAL MILIEU   62

    SUMMARY   68

4 **PREDICTING PARTICIPATION** 71

COURSE PLANS 72

THE PREDICTORS 75
Background Variables 75
Ability 76
Attitudes 78

DEALING WITH LONGITUDINAL DATA 81

RESULTS FROM PREDICTING COURSE PLANS 83
Middle School Students 85
High School Students 92

SUMMARY 95

5 **SUGGESTIONS FOR CHANGES** 97

CHANGING THE MATHEMATICS CLASSROOM 101

DESIGNING SPECIAL PROGRAMS FOR STUDENTS
ON CAREERS IN MATHEMATICS 107

DESIGNING PROGRAMS FOR PARENTS,
TEACHERS, AND GUIDANCE COUNSELORS 108

INTRODUCING NEW MATHEMATICS COURSES 108

REQUIRING FOUR YEARS OF HIGH SCHOOL
MATHEMATICS 111

CONCLUSION 112

BIBLIOGRAPHY 115

APPENDIX A   TENTH-GRADE QUESTIONNAIRE 131

APPENDIX B   ADDITIONAL TABLES 147

INDEX 159

List of Figures

FIGURE 1.1   Sex Differences in Enrollments in High School Math
              Courses   4
FIGURE 1.2   Plans for Participation in Mathematics   6
FIGURE 1.3   Spatial Abilities Tests   11
FIGURE 3.1   The Attitude Factors of Like/Dislike and Easy/Difficult   44
FIGURE 3.2   The Attitude Factors of Enjoyable/Anxiety-Provoking
              and Creative/Dull   45
FIGURE 3.3   The Pragmatic Factor of Usefulness   55
FIGURE 3.4   Students' Career Aspirations   59
FIGURE 3.5   The Social Factors of Encouragement from Others and
              Sex-Typing of Mathematics   63
FIGURE 3.6   Distance Between Self-Concept and Stereotype of
              Mathematician   64
FIGURE 4.1   Course Plans in Mathematics   72
FIGURE 4.2   Dividing Longitudinal Data into Component Parts   82

List of Tables

TABLE 2.1   Composition of the Sample by Grade and Year of Testing   24

TABLE 2.2   Description of Sample   26

TABLE 2.3   Reliabilities of Attitude Scales, Primary Sample   34

TABLE 3.1   Number of Students in Each Sex-Age Group   43

TABLE 3.2   F-Ratios for Main Effects and Interactions for Cross-Sectional Analyses of Mathematics Attitude Scales   46

TABLE 3.3   F-Ratios for Main Effects and Interactions for Longitudinal Analyses   47

TABLE 3.4   Reasons Students Give for Liking or Disliking Mathematics   49

TABLE 3.5   Topics in Mathematics that Students Especially Like or Dislike   50

TABLE 3.6   Students' Suggestions for Changing the Teaching of Mathematics   51

TABLE 4.1   F-Ratios for Main Effects and Interactions from the Analyses of Course Plans   74

TABLE 4.2   Means on Ability Measures   76

TABLE 4.3   Factor Loadings for Items in the Feeling Subscales   79

TABLE 4.4   Prediction Results Using Two Years of Data   85

TABLE 4.5   Separate Prediction Results for Males and Females Using Two Years of Data   87

TABLE 4.6   Prediction Results Using Three Years of Data   88

TABLE 4.7   Separate Prediction Results for Males and Females Using Three Years of Data   90

TABLE 5.1   Relative Rankings of Liking of School Subjects   103

Preface

The research reported in this book had as its initial goal coming to understand why so few women went on to study advanced mathematics. The Spencer Foundation found such a goal worthy of support and funded the first year of the project (1976–1977). The National Institute of Education shared an interest in this issue and funded the second and third years under contract #400-77-0099. As the research project progressed, it became clear that dropping out of mathematics was not a totally female phenomenon, but was happening among males as well. And for both sexes the reasons for not electing optional mathematics courses in high school were shortsighted—"I'll never need it"—and sometimes patently false —"I'm no good in math," a remark made by some highly able students. To understand how to steer students back into mathematics or encourage continued participation, I have looked at why they do not wish to continue, and have suggested ways of reversing their reasons by convincing them through their own logic that they should continue.

There were two purposes for translating the research into a book: wanting to record the results of a detailed longitudinal study, and wanting to translate those results into concrete, positive suggestions for teachers of mathematics. I therefore set out to produce a book that was readable by teachers, yet still usable by researchers in psychology and education. For those among the audience who are not knowledgeable about statistics, I have tried to explain the meaning of the techniques employed and to suggest interpretations of the results. It is hoped that these discussions will prove useful to the person who always wanted to know what those p's, F's, and r's meant, and not boring to those people who already know.

In the course of reporting and discussing project results, I have said a great deal about how I believe mathematics should be taught. These beliefs are based on project findings, considerable experience teaching statistics to anxious psychology majors, and extensive discussions with teachers and students about their experience with mathematics. My point of view has been heavily influenced by the successful discovery learning I have seen in schools and this is clearly visible in some of the suggestions I make for improving mathematics teaching. However, I am aware of the problems with this approach, and try to look at the advantages and disadvantages of many strategies for teaching mathematics.

The production of the book has required substantial efforts from a group of committed people. As project director and senior author, I have had final responsibility for the manuscript, but three colleagues have contributed to the writing in major ways. Cynthia Char produced several drafts of Chapter Three on our way to understanding the material and learning to present it in the "best" way. Glenn Takata produced preliminary drafts of Chapter Four after having analyzed the complex of data on the computer. And Nancy Irwin filled the onerous role of editor for our work, weaving flowing prose into solid arguments. Several people have contributed to the project by generating ideas for new directions in the analysis or reviewing reports and suggesting new interpretations of findings: Wendy Abt, Dennis Affholter, Andy Anderson, Susan Curtin, Steven Fosburg, Jay Magidson, William Reddy, Robert Rosenthal, Richard Ruopp, Alice Schafer, Lucy Sells, Judy Singer, Linda Stebbins, and Fran Stubblefield. Many have done impressive work as research assistants and secretaries: Alwina Bennett, Tara Feraco, Samuel Gilmore, Wendy Goldberg, Aleen Grabow, Annie Hondrogen, Karen Keefe, Claudia Kelly, Ona Langer, Abigail Millikan, Sharon Pinkham, Andromache Sheehey, Judy Sprotzer, and Nancy Stevens. To each of them I extend many thanks for untiring cooperation and support.

Finally, I would like to express my sincere appreciation to the three school districts which allowed this research to happen. Many more staff and students than I could mention made me welcome and willingly filled out questionnaires and answered interview questions. They know who they are and that I fully appreciate the efforts they had to make to meet my research needs. Many thanks.

Lorelei Brush
Abt Associates Inc.
Cambridge, Mass.
October 1979

πΣπΣπΣ **1** πΣπΣπΣ

The Problem

PROLOGUE

From the conception of this study in 1976, its goals have been to trace the development of students' desire to participate in mathematics or to avoid it whenever possible, and to examine closely the differences between boys and girls in their reasons for enrollment or avoidance. The rationale behind such a study is that too many students, particularly girls, are dropping mathematics in high school as soon as they can. They do not realize that their lack of mathematical preparation will restrict their job opportunities, keeping them out of many of the higher-paying jobs and seriously limiting their opportunities for growth and advancement in other jobs. We need to know why students decide to quit mathematics, in order to reverse the process and encourage them to keep studying the subject.

To trace this decision-making process, we collected information from nearly 2,000 students in three New England school systems. In the 1976–77 school year these students were enrolled in the sixth, ninth, and twelfth grades. The two younger groups were followed through the 1977–78 and 1978–79 school years, so at the end of the study three years of longitudinal data were available on a very large number of young men and women.

We discussed a wide variety of topics with students in searching for the reasons they used to decide whether to take more courses in mathematics. We looked at the background factors of socioeconomic status and ability; students' attitudes toward mathematics as an easy, creative, enjoyable, and useful field; and students' interpretations of the clues they were given by their social milieu concerning the appropriateness of their studying mathematics. This meant asking about the degree of encouragement they felt they were receiving from teachers, parents, and peers; the degree to which they felt it was appropriate for females as well as males to study mathematics; and the degree to which they felt they shared any personality characteristics with mathematicians. Students completed a questionnaire on these topics each year. In addition, a number of girls were asked to talk at length about them in interviews.

This book will argue the case that there is a critical problem for education in the low female enrollment in advanced mathematics courses. We are not properly preparing students for the job market they will face on graduation from high school or for enrollment in college courses that are required for better-paying jobs for college graduates. This argument is presented in the remainder of Chapter One, along with a discussion of the possible reasons students avoid mathematics. Chapter Two describes the organization of the three-year study that addressed this problem. In Chapters Three and Four the results of the study are presented in detail, and in Chapter Five we talk about specific strategies that educators might use to encourage more students to enroll in advanced mathematics courses.

DESCRIPTION OF THE PROBLEM

One of the most startling pieces of information about women's lack of participation in mathematics came from Lucy Sells's *Fact*

Sheet on Women in Higher Education as she discussed applicants
for the class of 1976 of the University of California at Berkeley:

> A study of admission applications of Berkeley freshmen shows that
> while 57% of the boys had taken four years of high school math (first
> year algebra, geometry, second year algebra, trigonometry and solid
> geometry), only 8% of the girls had done so (Sells, 1974, p. 3).[1]

Why should there be such a large discrepancy between high
school boys and girls in mathematics enrollment? Genetic differences may account for a small part of this, but it may also be
possible to explain the differences totally on the grounds that
experiences with mathematics differentiate the sexes. Perhaps
teachers—and others—treat girls differently from boys when it
comes to mathematics. Perhaps girls and boys respond differently
to identical treatment in mathematics classes. Perhaps it is the
motivation to study mathematics that distinguishes one sex from
the other. Data as dramatic as Sells's demand that someone take a
closer look at this phenomenon and try to uncover its causes—and
that is the purpose of the study summarized in this book.

The Sells finding is by no means an isolated one. For example,
the Educational Testing Service (ETS) recently reported on the high
school mathematics preparation of students in the class of 1977
who took the Scholastic Aptitude Test (SAT) (Admissions Testing
Program of the College Entrance Examination Board, 1977). Sixty-
one percent of the boys taking the test had completed four or more
years of high school mathematics, compared to 40 percent of the
girls. Thus there may have been an increase in girls' participation in
mathematics between 1972, when Sells gathered her data, and
1977, but boys' participation rate was still far greater than girls'.

And this finding is not limited to high-ability groups who apply
to the University of California at Berkeley or take the SAT. A recent
study by the National Assessment of Educational Progress (NAEP)
showed that the same is true of high school students in general.[2]
During the 1977–78 school year, NAEP sampled 25,000 17 year olds
from high schools across the country. They found that significantly
more males than females enrolled in the more advanced high
school mathematics courses: trigonometry, computer science, and
precalculus or calculus (see Figure 1.1), showing that, in general,
high school girls across the country were not taking as many mathematics courses as boys.

Unfortunately, there is also some evidence that even very
young girls and boys differ in their desire to take mathematics. For
instance, in the present study we found sex differences in *plans* to

FIGURE 1.1.
Sex Differences in Enrollments in High School Math Courses[a]
(N = 25,000)

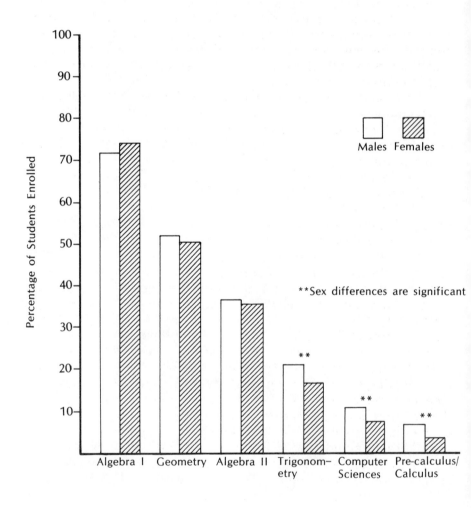

[a]Data from National Assessment of Educational Progress, Second Mathematics Assessment, 1977–78, courtesy of Dr. Jane Armstrong.

study mathematics as early as seventh grade, as well as differences in actual course enrollment in high school. Figure 1.2 shows the mean number of times sixth to eighth graders said they would choose to take mathematics over a course in the humanities (English, history, or foreign languages), as well as the mean number of mathematics courses which ninth to twelfth graders planned to take (or had taken) in high school, beyond the two years required.

In the seventh grade, male students demonstrated a much stronger preference for mathematics courses than did female students; in the tenth and eleventh grades male students planned to take (and in twelfth grade had actually enrolled in) more mathematics courses than females.

These limitations take effect for some students through a restricted choice of college majors. At the University of California at Berkeley, for example, four years of high school mathematics is required for admission to introductory calculus. Calculus, in turn, is required for a major in every field except the stereotypically female domains—humanities, social sciences, library science, social welfare, and education. As Sells (1974) has stated, high school mathematics is serving as a "critical filter" for entry into college departments, restricting women to what have proven to be their traditional fields.

Moreover, this pattern is repeated at higher levels of the academic world. Women are underrepresented not only in optional high school math classes, but also among tenure-track faculty in Ph.D.-granting departments of mathematics. (American Mathematical Society, 1975; Centra, 1974; Schwandt and Kreinberg, 1979). In the McCarthy and Wolfle (1975) account of doctorates granted to women from 1969 to 1975, it is stated that women obtained less than 10 percent of the Ph.D.s granted in astronomy, economics, mathematics, computer science, applied mathematics, physics, engineering, and operations research, among other fields. In no field that requires quantitative study did women receive over 30 percent of the Ph.D.s granted. The fields in which women were most frequently represented (that is, received *more* than 30 percent of the Ph.D.s granted) were home economics, art history, Romance languages, comparative literature, social work, health sciences, English, psychology, anthropology, and library science.

The importance of all these facts and figures should be clear: it is desirable for students of either sex to involve themselves in so much mathematics because the earlier students stop taking mathematics, the more severely they limit their choice of occupations and their chances to earn an economically satisfying living.

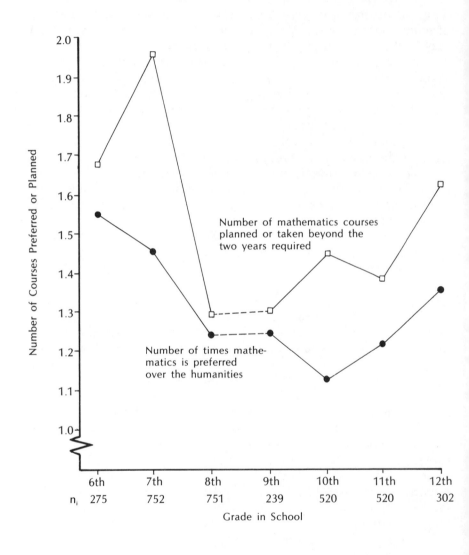

FIGURE 1.2.
Plans for Participation in Mathematics

□Males
•Females

When students enter the job market—whether directly from high school or after pursuing a higher education—similar limitations come into play. A look at the U.S. Department of Labor's *Occupational Outlook Handbook* (1976) brings this point home. This publication compares the projected number of job openings over the next 10 years in a large number of categories to the projected number of qualified applicants for those jobs. For most jobs that require a facility with numbers or explicit knowledge of mathematics, the prospects for employment are good, (although teaching is an exception). But for jobs that require training in English or history or even library science, the prognosis is grim. Job applicants—male or female—in the stereotypically female fields will find it difficult indeed to obtain satisfactory employment. They will face an ever-shrinking demand for their skills and increasing competition for every available job.

In addition, jobs which require quantitative skills pay higher salaries than those which require no such skills. For example, the College Placement Council (1978) recently reviewed starting salaries offered to college graduates, dividing these students by field of undergraduate major. The average hourly rate offered majors in engineering in 1978 was $7.99, an average annual starting salary of $16,668. For majors in the sciences the beginning hourly rate was $6.55; for those in business and management, $6.07; for those in economics, $5.46; for students in other social sciences, $4.84; and for those in the humanities, $4.77. Humanities graduates were thus offered an average annual salary of only $9,948. The difference in the hourly rate for humanities and engineering majors was a whopping $3.22, and the yearly salary difference was $6,720. Students of the humanities will find it difficult not only to locate a job, but also to earn a sufficient wage.

Finally, we must consider the issue of occupational flexibility, for it seems clear that misconceptions about the typical progress of careers may lead students to narrow their options early in high school. Often the emphasis in guidance offices is placed on "choosing a career," with a career portrayed as a single path a student will follow throughout adult life. But, in fact, relatively few adults these days follow a single career path. More and more adults are finding that the career they chose in high school or college is not sufficiently rewarding, that they have developed new interests, that their chosen occupation leads naturally into another very different one, or that circumstances present unlooked-for opportunities. Increasingly, when presented with career options they never considered in high school or college, people are changing careers in their late twenties or thirties or forties. And in many cases, what enables them to make such changes is a broad educational preparation that includes some familiarity with mathematics.

Students who have been able to avoid mathematics have also cut themselves off from many career opportunities; they have lost their flexibility and with it their marketability. They have also restricted the range of jobs in which they can seek personal gratification. We must convince our teenagers that mathematics is an essential career tool, whatever their present plans. Job options that require quantitative skills, such as computer work or business, should be kept open, as moves into these fields may become attractive alternatives later on. In addition, those who are comfortable with quantitative notions generally will find more open doors even in fields not directly concerned with mathematics.

What about those girls who feel that they do not need to worry about employment because they plan to be wives and mothers, supported by their husbands? Though many of these young women will indeed marry and have children, it is not likely that they will remain permanently out of the labor force. Many couples soon find that a second income is required for survival in today's economy, or at least is needed to purchase the amenities they desire. In 1977, 39 percent of mothers in intact families with children under 6 years of age were members of the labor force and 56 percent of mothers of children from 6 to 17 worked outside the home (U.S. Department of Labor, 1977). Moreover, the burden of supporting a family often falls on a mother. Husband and wife may seek a divorce, or the husband may die or be disabled, forcing the woman either to work or subsist at a much lower standard of living. As one might expect, most divorced women choose to work — 66 percent of those with children under 6, and 82 percent of those with children from 6 to 17 years of age, according to the U.S. Department of Labor (1977). Most women spend a significant part of their lives working outside the home, and young women need to be prepared for this eventuality.

There are obviously other, educationally valid reasons to encourage students to pursue mathematics, besides their need for such knowledge to find good jobs. The study of mathematics trains the mind in rational thinking, in reasoning from premises to a logical conclusion. The skills and understanding required to master mathematical processes are useful in everyday life for such simple feats as making change and also for more complex efforts, such as evaluating graphs in newspapers and being certain about what they mean. Moreover, we are living in an age of increasing technology, where an educated person is soon to be characterized as one who is knowledgeable about computers as well as more traditional fields. Thus, everyone needs to be able to adopt a mathematical perspective to examine today's world, and this need will grow in the future.

For many reasons, then, encouraging students to continue studying mathematics as long as they can is important. But how are we to achieve this goal? The practical advantages of a familiarity with mathematics, as set forth above, can be presented to students by guidance counselors, parents, and teachers. And with students who enjoy mathematics—the disciplined thought, the challenge of a difficult problem, the mastery of a tricky proof—we can discuss the many intrinsic rewards. But those students who do *not* find such rewards in problem solving, who may in fact systematically avoid the study of mathematics, pose a challenge to the educator. If we are to intervene effectively to overcome students' reluctance to enroll in mathematics courses, we must understand the causes of their aversion.

EXPLANATIONS OF AVOIDANCE

A number of scholarly papers and books have addressed the problem of mathematics avoidance, and four general types of explanations for students' actions have been presented: lack of ability; negative attitudes toward the subject; perceived lack of usefulness of mathematics in future life; and a discouraging social milieu.

Lack of Ability

Is there a relationship between students' mathematical ability and their pursuit of mathematics, that is, their enrollment in mathematics courses? Many students feel that they simply do not have the necessary aptitude, and they choose not to take any more mathematics courses because they feel they cannot handle the material.

Although such self-perceptions are sometimes accurate, we must reexamine their implications. It is probably true, for instance, that calculus is not accessible to everyone; the level of abstraction required to understand this branch of mathematics is too high for some. But this does not mean that these students' mathematics curricula must end with arithmetic. Most schools have a lengthy business mathematics sequence, including bookkeeping and accounting courses; and some schools offer computer programming or some variant of "essential mathematics" or "consumer mathematics." Many careers in which job prospects are good require just such skills. Avoiding calculus does not have to mean avoiding high school mathematics. The taint that entering a business mathematics course may have on a college-bound student could also discourage some students. If so, this issue must be dealt with in high schools. But some form of mathematics can be beneficial to students in every grade if the importance and appropriateness of this science are made clear.

Teachers must also bear in mind that students' perceptions of their own mathematical ability may not be accurate. Parsons (1979), for example, has found that girls consistently state that mathematics is harder than do boys, even when girls and boys maintain the same grades and to outward appearances girls are not having any greater difficulty with the subject.

What, then, do research findings show about actual differences in mathematical ability—including sex differences, if any—and about the relationship between ability and enrollment in mathematics courses? The usual measures of mathematical ability are measures of "aptitude" or "achievement"—tests with subscales for manipulation of concepts and for computations. These measures do show some sex differences and may predict enrollment (Aiken, 1970a, 1971, 1976; Anastasi, 1958; Astin, 1974; Backman, 1972; Fennema, 1974a; Fennema and Sherman, 1977; Flanagan, 1976; Flanagan et al., 1964; Fox, 1975b; Fox, Fennema, and Sherman, 1977; Garai and Scheinfeld, 1968; Glennon and Callahan, 1968; Hilton and Berglund, 1974; Mullis, 1975; Sheehan, 1968; Wilson, 1972). Maccoby and Jacklin conducted an extensive literature review in 1974 and concluded that boys and girls do not show differences in mathematical ability in elementary school, but that differences begin to appear when children reach puberty, and are found more frequently with increasing age. That is, when differences between the sexes in mathematical ability are found, teenaged boys score higher than girls. Sherman (1978) has recently reexamined the Maccoby and Jacklin findings and has some disagreements with them, but the general tenor of the findings still holds.

Another group of tests measures "spatial" ability. This capacity to manipulate objects in the head is held by some researchers to be closely related to mathematical ability and necessary for certain kinds of mathematical achievement (for example, geometrically based mathematics). Tests of spatial skills may require that the stu-

FIGURE 1.3.
Spatial Abilities Tests[a]

A. Gestalt Completion Test: Look at the incomplete picture and try to see what it is.

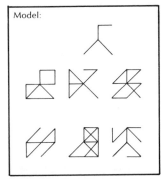

Model:

B. Hidden Patterns Test: Identify all complex figures which contain the model.

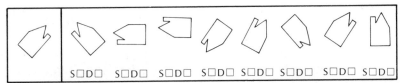

S☐D☐   S☐D☐   S☐D☐   S☐D☐   S☐D☐   S☐D☐   S☐D☐   S☐D☐

C. Card Rotations Test: Decide whether each of the cards on the right is the same as (S) or different from (D) the card on the left. Cards may be slid around, but cannot be lifted off the page to be considered the same as the card on the left.

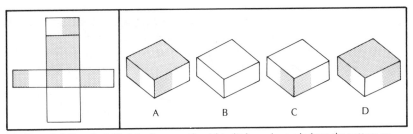

A   B   C   D

D. Space Relations Test: Find the figure on the right which can be made from the pattern on the left.

[2]Tests A, B, and C copyright © 1976 by Educational Testing Service, Princeton, N.J., 08541. All rights reserved. Reproduced from the *Manual for Kit of Factor-Referenced Cognitive Tests*, R. B. Ekstrom et al. Test D copyright © 1972 by The Psychological Corporation, 757 Third Avenue, New York, N.Y., 10017. Reproduced by special permission from the *Differential Aptitude Tests*, G. K. Bennett, H. G. Seashore, and A. G. Wesman.

dent identify an object from a suggestive picture that has eliminated parts of the object (see Figure 1.3A), pick out a simple figure when it is embedded in a more complicated design (see Figure 1.3B), identify an object when it has been rotated in the same plane (see Figure 1.3C), or match an object to its prototype when the prototype is two-dimensional and the object three-dimensional (see Figure 1.3D). In studies of the most frequently used test, composed of tasks like that in Figure 1.3D (the space relations section of the Differential Aptitudes Tests[3]), sex differences have been found, indicating that boys are more able to handle spatial tasks than are girls (Aiken, 1971, 1973; Fennema, 1975; Fennema and Sherman, 1977; Sherman, 1967). It may be, then, that differential spatial abilities between the sexes contribute to differential mathematics enrollments.

In sum, the evidence from aptitude, achievement, and spatial skills tests indicates that sex differences may exist, and that ability may contribute to students' desire to enroll in mathematics courses. In our study we look closely at the effects of such ability factors.

Negative Attitudes Toward Mathematics

It is a reasonable hypothesis that students' attitudes toward a subject affect their desire to participate and their actual choice of courses. If a student really likes mathematics, thinks it is easy and fun, he or she will take it. But if a student dislikes the subject, thinks it is difficult, or is afraid of appearing foolish in class, then that student will probably stop taking mathematics as soon as possible.

Some researchers have tapped attitudes toward mathematics by asking a series of general questions with which students are asked to agree or disagree, and computing a summary score ranging from very negative to very positive. Wilbur Dutton, for example, includes the following group of items in his scale: "I have never liked arithmetic"; "Sometimes I enjoy the challenge presented by an arithmetic problem"; "Arithmetic is very interesting"; "I never get tired of working with numbers." Students are asked to check all items that apply. Dutton (1951, 1954, 1956, 1962, 1965, 1968; Dutton and Blum, 1968) has attached a value (from 1.0, for the most negative attitude, to 10.5, for the most positive attitude) to each item, and scores each student by computing the mean value for the items checked by that student (Abrego, 1966; Adams and Von Brock, 1967; Aiken, 1963, 1972a; Aiken and Dreger, 1961; Anttonen, 1969; Bassham, Murphy, and Murphy, 1964; Callahan, 1971; Campbell and Schoen, 1977; Carey, 1958; Degnan, 1967; Duckworth and Entwistle, 1974; Dutton, 1951, 1954, 1956, 1962, 1965, 1968; Dutton and Blum, 1968; Fedon, 1958; Fellows, 1973; Gilbert, 1977; Gilbert and Cooper, 1976; Holly, Purl, Dawson, and

Michael, 1973; Hungerman, 1967; Hunkler, 1977; Hunkler and Quast, 1972; Lyda and Morse, 1963; McCallon and Brown, 1971; Muscio, 1962; Reys and Delon, 1968; Roberts, 1969; Smith, 1964; Stright, 1960; Todd, 1966).

Other researchers have tried to be much more specific in investigating attitudes toward mathematics, suggesting that different topics arouse different levels of feeling (Aiken, 1974; Bachman, 1970; Fennema and Sherman, 1977; Husen, 1967; Sherman and Fennema, 1977).[4] For example, separate scales exist to measure a respondent's perceptions of the difficulty of mathematics, the degree to which mathematics is fun as opposed to boring and monotonous, and the degree to which a respondent can use his or her own ideas in studying mathematics as opposed to merely memorizing what others have already figured out. There are also scales to measure self-confidence in performing arithmetic operations; the student's attitude toward being successful in mathematics; and, probably the strongest attitude of all, "anxiety" about mathematics.

Mathematics anxiety (or number anxiety, or "mathophobia") may be defined as an extreme negative reaction in the face of mathematics classes and assignments which results in discomfort and may also cause restlessness and irritability. It is a reaction which takes dislike a step further, adding a suggestion of fear. Many people experience some annoyance with an assignment that is too long or too boring, but it is one more step to *anxiety*—when the student feels sure that he or she cannot do mathematics and does not want to be anywhere near the places where mathematics is done.

Most people can think of at least one experience with mathematics which caused them anxiety. For example, one of this book's contributors vividly recalls taking a high school mathematics test which had instructions stating that all answers were to be written in equation form. She couldn't remember to save her soul what an equation was, so she was certain to fail the exam before working even one problem. Although this was a very anxiety-provoking situation, it was an isolated instance for this person. For some others, such instances have occurred with such regularity that mathematics anxiety has become the normal state of affairs.

Take Doug, for example, a student who was sick off and on for most of one school year. Each time he came back to school he had to fight harder to catch up in mathematics than in other subjects. After a somewhat protracted illness, he was sure he never could catch up, and he couldn't recover any sort of enjoyment of mathematics. The days he missed haunted all of his mathematics work; even if he understood today's assignment, there was the nagging fear that tomorrow's would require knowledge that he had missed. His anxiety toward the subject was firmly established and never disappeared.

And there is the poignant story of Sandy and math baseball. It appears that Sandy's fears about mathematics began in the third grade when her teacher's favorite mathematics game was a variant on the spelling bee called math baseball. Two captains were appointed from the class, and they chose teams for the game. The teams lined up on either side of the room, lead-off batters closest to the blackboard. The teacher "pitched" a problem and the lead-off batters ran to the blackboard and solved it as fast as they could. The one who came to the correct solution in the shortest time scored a run.

Sandy wasn't a fast runner, so she was usually at a disadvantage before she even started the calculation. And although she was very accurate in doing arithmetic calculations, Sandy was slower than a lot of other children. So she could never score a run for her team. Even though other pupils didn't say anything to her, Sandy knew she was a liability to her friends, and she grew to dread these episodes of public problem solving. Sandy never recovered her initial enjoyment of mathematics. She couldn't take comfort in the fact that she got the right answers and did well on tests, because in the most important arena—respect among friends—she didn't feel successful at all.

These examples demonstrate what anxiety is like for the student and how it may begin. Research into situations which cause anxiety shows that interpersonal or public situations in which mathematical abilities are being evaluated generate the most anxiety (Brush, 1978). Such situations as taking a mathematics test, being watched while adding up a column of figures, and serving as treasurer of a club all give others the opportunity to judge a student's skill in mathematics, and all are reported by high school and college students to be anxiety-provoking. On the other hand, doing simple calculations like addition and division—alone and without time pressure—is not bothersome to most people.

Research findings also suggest that students who are more anxious about doing quantitative work do not enroll in optional mathematics courses (Biggs, 1959; Brush, 1978; Sherman and Fennema, 1977). And students with a relatively high level of anxiety in mathematics tend to have other negative attitudes toward the subject (Brush, 1978; Fennema and Sherman, 1976). They tend to lack confidence in their ability and think of the subject as difficult. They frequently report that they not only dislike mathematics, but find it boring or monotonous. They feel it is an exercise in memorizing what everybody else already knows and, therefore, is in no sense a challenge.

Of course, not every student holds negative attitudes toward mathematics. Some clearly like mathematics and see it as interesting, challenging, fun, open to new ideas, and easy if you work at it. For the purpose of the study reviewed in this book, we would like to find out who these students are and how they differ from those who dislike the subject. Several studies have found that girls more often hold negative views about mathematics than boys (Aiken, 1972a, 1976; Aiken and Dreger, 1961; Carey, 1958; Dreger and Aiken, 1957; Dutton, 1956; Fennema and Sherman, 1977; Hilton and Berglund, 1974; Husen, 1967). Throughout elementary school, the sexes rarely differ in their attitudes, but after the age of 12 any sex differences which are found show that girls dislike mathematics more than boys.[5] This is of particular concern in the present study.

In sum, research findings and anecdotal evidence suggest that many kinds of conditions may contribute to a general like or dislike of mathematics, that girls tend to have more negative attitudes toward mathematics than boys, and that negative attitudes tend to accompany failure to participate in optional or advanced mathematics courses.

Lack of Usefulness of Mathematics for Future Life

It is logical that students who see no application for mathematics in their own future careers or personal lives are less likely to take optional or advanced courses in the subject. This attitude toward mathematics may accompany the sort of negative attitudes discussed in the preceding section, but it may also stand alone as a legitimate, pragmatic judgment. Able students of mathematics who are also confident of their ability and enjoy the subject may decide to discontinue their studies because, in their judgment, further study is not gainful. Career choices, other talents and interests, or personal plans may cause students to feel that the time needed to pursue mathematics is better spent elsewhere. Clearly, for some people such a decision does not have the negative overtones of "opting out." But even in these situations educators may want to ensure that such a decision is not made prematurely.

Lynn Fox provides some interesting evidence on this point in her discussion of seventh-grade girls who participated in statewide mathematics contests in the early 1970s (Fox, 1975a, 1976a; Fox and Denham, 1974). Here we have a group of very talented girls who have a demonstrated ability in mathematics far above average. Fox asked them about their career plans in the context of discussing enrollment in accelerated mathematics classes. Girls who said that their future plans included marriage and family were far less likely to want to join an accelerated mathematics class than were girls with other career plans. Even those girls (and boys) who couldn't specify any particular career but said they were sure they would have one were more likely to want to enroll in a special course. It appears that students don't have to know *exactly* what they want to do or feel that mathematics is an integral part of their anticipated careers to say they want to take advanced mathematics courses. If they know that they will be doing *something* outside of the home, they are more likely to decide to keep their options open with regard to mathematics.

In the late 1970s it is difficult to find a group of girls or young women who will state flatly that marriage and family are their *only* future concerns. They may believe that they need not keep job options open, but they won't say that on questionnaires. So instead of relying on one question which asks about career plans, researchers have resorted to the use of scales to assess sex differences in opinions about the value and usefulness of mathematics (Aiken, 1974; Armstrong, 1979; Fennema and Sherman, 1977; Haven, 1971; Hilton and Berglund, 1974; Husen, 1967; Sherman and Fennema, 1977; Wise, 1978). They ask for agreement/disagreement with such statements as, "An understanding of mathematics is needed by artists and writers as well as scientists"; "I will use mathematics in many ways as an adult"; "It is important to know mathematics in order to get a good job." They have sometimes found sex differences in the perceived usefulness of mathematics, but not universally. For example, in only two of four high schools in the Fennema and Sherman (1977) study did males see mathematics as more useful than females did and in the Aiken (1974) sample of college freshmen no sex differences were apparent. When investigating the relationship of perceived usefulness to participation in optional courses, researchers usually find a correlation suggesting that students who find mathematics more useful are more likely to enroll (Armstrong, 1979; Haven, 1971; Sherman and Fennema, 1977).

Thus, the question of the usefulness of mathematics seems worth pursuing. There is a suggestion of sex differences in perceived usefulness, and evidence that perceived usefulness is related to participation.

A Discouraging Social Milieu

Finally, various social influences may come into play in students' decisions to participate in or withdraw from mathematics courses. That is, students are likely to enroll in courses in which they feel they belong, and avoid those about which they feel uncomfortable. What are the sources of such feelings? Students may be influenced by stereotypes they hold of persons involved in mathematics as opposed to, say, the humanities. Not only may such stereotypes dictate to students the *appropriateness* of pursuing mathematics—for example, a seventh-grade girl might think mathematics is only for boys—but a student's perception of his or her own personality characteristics may or may not jibe with those of the stereotypical mathematician or artist. In addition, students' feelings about studying mathematics are affected by the support—or the lack of it—that they receive from those around them, whether teachers, parents, or peers.

What stereotype of a mathematician do high school students hold? Clearly, one important factor could be the stereotyping of mathematics as a male domain. The data on participation in mathematics in high school and beyond, presented at the beginning of this chapter, demonstrate that more males are taking advanced mathematics courses than females. It may well be that high school girls notice this fact and feel odd about continuing to enroll in classes dominated by boys. Those women who once took physics with 2 or 3 females in a class of 28 have mixed memories of the experience. The experiments conducted with three male lab partners had a certain pleasant excitement, but how much of this came from being "allowed" to take part in men's work and the men's world?

How, then, do girls interpret these experiences in male-dominated classes? Do they believe that it is more appropriate for boys to study advanced mathematics? Is the field indeed stereotyped as a male domain? Is there something inherent in the subject matter of mathematics which makes it more appropriate for men than women? Research evidence on this point suggests that many people do consider mathematics as more appropriate for men, and that this sex typing can inhibit females' achievement and their desire for continued participation in the field. (Aiken, 1970b; Casserly, 1975; Ernest, 1976; Fox, 1976a, 1976b; Mokros and Koff, 1978; Sherman and Fennema, 1977; Stein and Smithells, 1969).

Another potential social influence is the fit between one's personality and the perceived characteristics of successful mathematicians. Is there a certain kind of personality needed for success in mathematics, especially to pursue advanced studies in the field? If students compare themselves to that personality and find themselves different, do they stop taking mathematics? Such influences, of course, may form part of a conscious decision, but they may also color students' perceptions in less obvious ways.

These questions have been examined by eliciting students' perceptions of mathematicians and their perceptions of their own personalities, and then comparing the stereotypes and self-images which emerge. Results indicate that both male and female students see mathematicians as rational, wise, responsible, cautious, and stable but lacking in sensitivity, gentleness, warmth, and sociability (Ahlgren and Walberg, 1973; Beardslee and O'Dowd, 1961; Boswell, 1979; Brush, 1979; Fox, 1976a; Hudson, 1967b; McNarry and O'Farrell, 1971). Mathematicians have some attributes that are stereotypically masculine, but no stereotypically feminine ones (Bem, 1974; Rosenkrantz, Vogel, Bee, Broverman, and Broverman, 1968). The distance between students' self-images and their perceptions of mathematicians is sometimes significantly less for males than for females (Brush, 1979). Although this distance variable has not been shown to be a strong *predictor* of students' enrollment in mathematics courses, it was shown to be correlated with participation. Students who champion the rational sides of themselves and downplay sensitivity do seem to enroll more frequently in advanced mathematics courses (Brush, 1979).

Finally, does the degree of support students feel they receive from their social environment affect their enrollment in mathematics? What sort of feedback — positive or negative — do students get from their parents, teachers, and peers? Research findings sometimes indicate that parents and teachers (and, once in a while, even peers) think that boys will outperform girls in quantitative subjects in high school, and that each of these social groups seems to encourage boys in their pursuit of mathematics courses more than girls (Casserly, 1975; Ernest, 1976; Fennema and Sherman, 1977; Kaminski et al., 1976; Sells, 1974; Sherman and Fennema, 1977). Since we know that encouragement is an effective motivation for participation (Anderson, 1963; Casserly, 1975; Ernest, 1976;

Haven, 1971; Poffenberger and Norton, 1956; Rosenthal and Jacobson, 1968), these sex differences in expectations and encouragement would seem a powerful predictor of differential enrollment.

In sum, various sorts of social influences appear to stereotype mathematics as a male domain. The consequences of such a stereotype may be bad for boys as well as for girls. If girls' options are narrowed by discrimination, so, too, are boys'; they may *have* to be good in mathematics to be masculine. What happens to the teenage boy who is not very good in mathematics? One young man known to another of this book's contributors was very wary of taking mathematics. After relating various episodes in his mathematics career, Jeff asked if we wanted to know the truth about why he was so scared of mathematics. He then asked that the tape recorder be turned off — he didn't want anyone else to hear the reason. Finally, he confessed that he knew that mathematics was a masculine subject, or at least that males were supposed to do well in mathematics. And he usually did do well. But every day in mathematics class he was overcome by the idea that if he failed, he couldn't call himself a man anymore. Because he couldn't face the constant threat of losing his masculinity, Jeff refused to take any more mathematics. The anxiety was too great.

It seems, then, that neither sex "wins" when a masculine character is attributed to mathematics. The explicit consequences of this stereotype need to be examined closely to determine the strength of their impact on all students' enrollment.

SUMMARY

The research literature indicates that women are entering mathematics courses less often than men, and this difference in participation has been variously explained by lower mean scores on mathematical ability among women, more negative attitudes, perception of mathematics as less useful, and discouraging influences of the social milieu. We have seen that sex differences are sometimes found in each of these four reasons for avoiding mathematics, and that each reason may be related to lack of participation. We turn now to a description of the research that this book summarizes. Each of the potential reasons for lower female participation will be subjected to close scrutiny and evaluated against each of the others to determine the causes of mathematics avoidance in all students and to isolate the particularly potent causes for girls.

NOTES

1. These percentages were derived from looking at a random sample of student applications to Berkeley in 1972. Applications from one hundred males and one hundred females were checked. The extreme differences between the sexes motivated a second sampling of the same application pool, and this second selection showed 57 percent of the men and 33 percent of the women to have four years of high school mathematics (Sells, personal communication). Since the results of the first sample received wide publicity, they are reported here, with recognition of the fact that they represent a wider divergence between the sexes than may, in fact, have been the case.
2. Unpublished data from the National Assessment of Educational Progress, Second Mathematics Assessment, 1977–78. For additional information, contact Dr. Jane Armstrong, Education Commission of the States, National Assessment of Educational Progress, Suite 700, Lincoln St., Denver, Colorado 90295.
3. The Differential Aptitude Test is available from The Psychological Corporation, 757 Third Avenue, New York, N.Y. 10017.
4. Mathematics anxiety, in particular, is discussed in Biggs, 1959; Brush, 1978; Dreger and Aiken, 1957; Gough, 1954; Lazarus, 1974, 1975; Richardson and Suinn, 1972, 1973; Sepie and Keeling, 1978; Suinn, Edie, Nicoletti, and Spinelli, 1972; and Tobias, 1978.
5. No sex differences were found in research by Abrego (1966), Callahan (1971), Dutton and Blum (1968), Hungerman (1967), and Roberts (1969), and elementary school girls were found to be more positive about arithmetic than boys in Stright (1960).

πΣπΣπΣ **2** πΣπΣπΣ

The Research Project

The purpose of our research project was to monitor the development of students' opinions about mathematics and their plans for involvement with the subject, trying to capture the moments when changes occurred and trace their causes. The three-year study started with children in the sixth, ninth, and twelfth grades. The sixth graders were followed through seventh and eighth grades, and the ninth graders through tenth and eleventh grades. An attempt was made to follow the twelfth graders by sending them postcards. However, after graduation only 30 percent returned the cards, and the data were not extensive enough to be used in a longitudinal study. Thus, at the end of the study, information was available on a range of grades from the sixth through the end of high school. The results of previous research indicated that there were few sex differences in ability in mathematics, desire to participate, or liking for the subject in elementary school, but that differences appeared in the junior high years. By starting with sixth graders, then, we could establish a baseline of boys' and girls' views toward mathematics and examine the changes that occurred over the years.

It is important in any project of this sort to make certain that a wide range of students is tested. If only very bright students or only low-income students are in the sample, for example, then conclusions can only be generalized to these groups. For this study we wanted the results to be meaningful for all students in this age range, so we chose to work in three very different school systems (albeit in only one part of the United States). One school district is a rural area supported by some farming and a great deal of work in nearby industries, the second is a small city with a strong white ethnic community, and the third is a suburb of a large city. As all three districts are in New England, there may be some concern about generalizing these results to other parts of the country. However, as we will explain, the students in the sample do represent a range of ethnic backgrounds and socioeconomic groups, so the results for these students should be applicable to a much wider group. Each reader may evaluate the similarity of his or her district to the ones used in this study in deciding the degree to which students are comparable.

We would also like to be certain that the results are not unique to a subset of students in a district, that is, that our sample is representative of the students in the districts. In most studies researchers choose a small group of students in a school to participate, assuming that this group represents the entire school population. In this study we have divided the students in each system into two groups and have compared descriptions of and results from the two to be sure that the sample on which complete data are available was not strangely biased. It was not, and we feel safe in stating that the results from our primary sample adequately reflect the attitudes and behavior of all students in these districts.

Three methods were used to gather information on students. The first was a questionnaire that asked students the same questions each year. The analysis of such an instrument allows one to look at the opinions and behaviors of all students in a particular grade (e.g., seventh graders in 1977–78), the changes in opinions or behaviors that occur from grade to grade both for the group as a whole and for individual students, and the differences in opinions or behaviors between different subsets of students (e.g., males and females).

The questionnaire also provided a vehicle for experimenting with new ideas. For example, in the first year of testing, students complained that they did not get a chance to express their feelings in prose, and that the rigid questions posed did not give them the opportunity to express their opinions fully. So on the second-year questionnaire three pages of essay questions were appended to solicit students' comments. Also, because analyses of the second-year data suggested that certain questions were not ideally stated, the third-year questionnaire contained two versions of some questions so that the effects of wording could be analyzed. Such flexibility in data collection proved to be very valuable in giving us support for the validity of our assertions.

The second method of data gathering was the interview. In the second and third years of the study we chose 48 girls to interview on a one-to-one basis to discover more about the experiences leading up to their present opinions about mathematics. In each year we selected the set of girls whose plans to study mathematics had changed the most from the preceding year. We talked with a group of students who liked mathematics much more this year than last and compared them with a group who disliked the subject much more now than in the past. This method allowed us to supplement the rather dry information from the questionnaire with a rich set of anecdotes about the process of learning mathematics.

For instance, we asked few questions about teachers on the questionnaire, limiting our inquiries to the degree of support that students felt they received from their teachers. In the interviews we asked about specific teaching techniques that students enjoyed and found useful, specific experiences in mathematics classes that were a pleasure or a disappointment, and particular characteristics of teachers that led to greater or lesser success with students. These tales gave us a much broader picture of mathematics as it is taught today, and a better perspective from which to evaluate all of the other information.

The third method was the collection of test scores from student records. We considered giving IQ and achievement tests to students, but the time that would have been required was prohibitive and the school systems' administrators were not enthusiastic about the testing, because they already required students to take such tests once a year. So with parental permission we collected IQ scores and mathematics, verbal, and spatial achievement scores from student records. These indices of ability could then be examined to see if they contributed greatly to the students' wish to continue studying mathematics.

In the presentation of results in subsequent chapters we have interwoven the questionnaire, interview, and ability results to present as complete a picture as possible of students' opinions about mathematics, how their opinions change with the age of the students, and how all of these factors contribute to students' plans to take or avoid more mathematics courses.

In the remainder of this chapter we describe the group of students who were included in the study, the scales used on the questionnaires, the essay questions which were added to the second-year questionnaire, the interview protocols from the second and third years of the project, and the standardized ability tests previously administered to students.

THE STUDENTS

Table 2.1 presents the composition of the sample by grade across the three years of the study. Students in the primary sample were selected from the sixth, ninth, and twelfth grades in each school system in the 1976–77 academic year. The sixth graders were then followed through the seventh and eighth grades, and the ninth graders through the tenth and eleventh grades. The twelfth graders were given the questionnaire only in the first year of the study. A postcard follow-up of these graduates was attempted in the winter of 1977–78, but only about 30 percent of the students responded — such a small group that no data could be reliably used as a follow-up to the original testing. To check that the primary sample was representative of the three school systems, a comparison sample was tested in the second and third years of the study. This sample contained all of the students in each school system who were not in the primary sample but were in the same grades as the students in that sample.

TABLE 2.1

Composition of the Sample by Grade and Year of Testing

	1976–77	1977–78	1978–79
Primary sample	6th, 9th, 12th	7th, 10th	8th, 11th
Comparison sample	—	7th, 10th	8th, 11th

Table 2.2 presents personal and socioeconomic characteristics of the students in the study. Data are presented for each age group in the primary and comparison samples.

The first issue we will address is the degree to which the students are representative of students in middle schools and high schools in the United States.* The total column in Table 2.2 provides some important information on this question. First, the percentage of minority students in this sample is lower than the national average. In the 1978–79 testing only 8 percent of the eighth- and eleventh-grade students in both the primary and comparison samples were minority students. According to the 1975 assessment by the Bureau of the Census, 13 percent of the population and 17 percent of the population under 15 years of age in the United States were members of ethnic minorities (U.S. Bureau of the Census, 1977). Thus, this sample includes fewer black, Native American, Spanish-surnamed, or Oriental children than one would expect in a sampling of children in the United States at large. Because most of the minority children in our sample were black, the mix within the minority group also differed from the general population.

The socioeconomic status of the students in our sample also diverges somewhat from that of the nation. To determine social class, we used the Hollingshead and Redlich (1958) scale of the occupation of the head-of-household. The working class category in Table 2.2 is a combination of the skilled, semiskilled, and unskilled labor categories on the Hollingshead and Redlich measure. Table 2.2 indicates that 46 percent of the eighth and eleventh graders in the primary sample, 38 percent of the students in the comparison sample, and 29 percent of the twelfth graders in the primary sample came from families in which the head-of-household had a working-class occupation. As the national average was 60 percent in 1975 (Women's Bureau), our sample may be seen as more privileged than the population of the country at large. The particularly high socioeconomic status of the twelfth-grade group may be due to two factors. First, there were work-study programs in two of the three school districts; we were not able to test these students, a group which probably included a disproportionate number of students from working-class families. Second, there is probably

---

*We will use the term middle school to refer to the sixth, seventh, and eighth grades, and the term high school to refer to the ninth through twelfth grades since this was the organization in each of the school systems we studied.

TABLE 2.2
Description of Sample

Primary Sample (8th and 11th Graders)

| Grade level (1978-79)[a] | | 8 | | | 11 | | |
School district[b]	A	B	C	A	B	C	Total
Number	91	95	89	90	70	79	514
Mean age (years)	13.07	13.25	13.23	16.08	16.26	16.19	
Sex (male/female)	41/50	63/32	43/46	35/55	26/44	31/48	239/275
Percent minority	0	12	8	2	17	9	8
Percent working class	67	57	15	66	68	6	46
Percent mothers employed	60	66	50	71	60	51	60
Percent college-bound	85	65	94	72	67	96	80

Comparison Sample

| Grade level (1978-79) | | 8 | | | 11 | | |
School district	A	B	C	A	B	C	Total
Number	76	87	78	26	62	55	384
Mean age (years)	13.12	13.06	13.17	16.12	16.19	16.22	
Sex (male/female)	38/38	44/43	34/44	15/11	28/34	34/21	193/191
Percent minority	0	14	10	0	8	11	8
Percent working class	49	56	6	52	67	7	39
Percent mothers employed	65	71	45	62	72	53	61
Percent college-bound	68	68	97	89	63	95	78

Primary Sample (12th Graders)

| Grade level (1976-77) | | 12 | | | | | |
School district	A	B	C				Total
Number	98	102	102				302
Mean age (years)	16.96	17.04	17.24				17.15
Sex (male/female)	60/38	43/59	40/62				143/159
Percent minority		—not assessed—					—
Percent working class	46	42	3				29
Percent mothers employed	55	60	35				50
Percent college-bound	66	79	98				82

[a]Background data were collected in all three years of the study. The most recent data are presented here because they apply to the sample in its final form, that is, after attrition had taken place.
[b]A is a rural school district, B a small city, C a suburb of a large city, all in New England.

a higher drop-out rate among students from working-class families; this would mean that fewer of these students remained in school, making them unavailable for our testing.

Nationally, an average of 56 percent of mothers of children from 6 to 17 work outside of the home (U.S. Department of Labor, 1977). Figures for our sample are very similar. Among the eighth and eleventh graders in the primary sample, 60 percent of the mothers are employed; in the comparison sample, 61 percent; and among the twelfth graders in the primary sample, 50 percent. It may be that fewer twelfth graders have working mothers because these data are older than those for the eighth and eleventh graders, or it may be that fewer students with working mothers appear in our sample because of a higher drop-out rate and a higher rate of participation in work-study programs.

The second question to ask about our sampling is whether the students in the primary sample represent their school systems as a whole or whether they are in some way a nonrandom subset. To answer this question we compared the members of the primary sample for each school district to the members of the comparison sample who were of comparable age. In school system A (the rural district), the two eighth-grade groups are significantly different with respect to the percentage of working-class families represented and the percentage of children planning to attend college.* The primary sample has both a larger proportion of students of working-class families and a larger proportion of students who plan to go to college. This is an interesting juxtaposition of differences; it seems that the primary sample may comprise more ambitious children than the comparison sample. The eleventh graders in school system A are almost all in the primary sample; thus the small differences which appear between samples may be due to the small size of the comparison sample. The only significant difference is in socioeconomic status; a larger proportion of the children in the primary than in the comparison sample are from families headed by skilled, semiskilled, or unskilled laborers.

In school system B (the small city), only one difference between the primary and comparison samples is significant. This occurs between the eighth-grade groups in the distribution of males and females. The primary sample has more males and fewer females than the comparison group. In fact, the groups for testing were chosen randomly, so that this difference does not serve as a proxy for ability or attitudes toward mathematics or any other factor that should disturb results; it just happened that the sex distribution for this group was biased toward males. The eleventh-grade samples from this school system do not differ significantly on any count.

In school system C (the suburban area), the two samples are also very similar. No differences between the two eighth-grade groups were significant, and only one was significant between the

---

*Throughout this book the word significant is used to mean that the probability of obtaining this result by *chance* is 5 percent or less (expressed as $p \leq .05$). That is, no more than 1 time in 20 would a result this extreme occur by chance; 19 times out of 20 this result occurs because of a meaningful difference between groups.

two eleventh-grade groups—again, a different distribution of the sexes. The primary sample has more females than males, and the comparison sample has just the reverse. As in school system B, this is not expected to affect the results in a substantive manner.

By and large, then, the students in the primary sample are representative of students in their school systems. The children in the primary and comparison samples from school systems B and C do not differ in any way that might be expected to affect the results of the study. And the only potential problems in school system A exist because the primary sample has a somewhat larger proportion of children from working-class backgrounds than the comparison sample and may have children with more ambitious plans for the future. The presence of a larger proportion of working-class children in the primary sample means that we should pay attention to this potential cause of differences in the results, watching for the interplay of social class with opinions about mathematics.

Finally, it should be noted that in any particular year of the study there were more children in these school systems than are in the research sample. Many children moved away in the course of the study, some were absent on a day of testing, and some were no longer in the same grade level as the children being tested; all available children were tested each year, but the primary and comparison samples contain only those students for whom reasonably complete data were available for all years of testing. We did, however, check to see if the children who were dropped from the sample were different from those who remained and found no differences of any significance. The results for our sample can therefore be taken as descriptive of students in these three districts.

THE INSTRUMENTS

In the fall of each year of the study all of the students in the appropriate grades were given a questionnaire. Testing was usually done in English classes because all students were required to take English the entire four years of high school. Filling out the questionnaire took from 20 to 45 minutes, depending on the reading speed of the students. For younger children in slower English classes the items were read aloud and the testing lasted an entire class period. For most of the older students there was time in class left over for the teacher to make announcements and discuss homework.

As was mentioned earlier, on the second-year questionnaire several essay questions were added so that students could express their opinions about mathematics in their own words.

In each year of the study, following data collection and some preliminary analyses, we selected and interviewed those girls whose desires or plans to take mathematics had changed the most from the preceding year. Half of these girls planned to take more mathematics; the other half planned to take much less.

In the first and second years of the project we collected ability scores after the completion of questionnaire testing in a school. In the first year we gathered scores for the children in the primary sample from their records, and in the second year we solicited scores for the comparison sample with the requirement that scores for both samples come from the same testing session.

We now turn to a detailed discussion of the data-collection instruments. We will first describe the questionnaire items, then the essay questions administered in the second year of the study, then the protocols for the second- and third-year interviews. Finally, we will describe the ability measures on which we collected students' scores.

The Questionnaire

The questionnaire for this study was very broad and covered many areas of potential interest: family background, course preferences and enrollment plans, expected liking for mathematics and anticipated grades in future courses, career plans, attitudes toward course work, and distance between students' self-perceptions and their perceptions of members of two professions (mathematician and writer). The expressed intention of this questionnaire was to assess students' views about mathematics. This goal was fulfilled, however, by looking at students' attitudes toward both mathematics and English. For example, if we had found only that a liking for mathematics declined over time, we would not have known whether a liking for all school subjects declined or only a liking for mathematics. By asking about English as well as mathematics, we could discern whether an attitude change is peculiar to one subject or more general.

In describing the questions asked of the students, we will talk about the kinds of questions asked, the format of the items, and the method of scoring. A copy of the tenth-grade questionnaire is included in Appendix A.

Family Background. The first questions concerned age, sex, ethnic background, school, father's and mother's occupations and education, and number and order of siblings. For analysis purposes, all non-Caucasian children were classed as minority, and father's occupation was used in rating the socioeconomic status of the family unless no father was present.

Course Plans. Questions in this area differed for middle school and high school students. The younger students were asked to circle the course they would prefer to take in each of nine pairs of courses. In each instance, one member of a pair was mathematics, physics, or chemistry, and the other English, social studies, or a foreign language. We tried to predict Preference for Mathematics from differences in students' backgrounds, abilities, and attitudes, and this preference variable was scored from 0 to 3 depending on the frequency of selection of mathematics. Preference for Humanities was scored as the number of instances humanities were preferred over mathematics. It is also possible to calculate a preference score for physics and chemistry, but because the focus of the present work is on mathematics, such a score will not be discussed.

For the older group, a list of the school's humanities and science departments was provided, and students were asked to write down the number of courses they intended to take in each department by the time of graduation. Course Plans in Mathematics, an extension of the younger students' Preference for Mathematics variable, was scored as the number of years beyond two (which was required in each high school) that the student planned to take or had taken. The scoring here ranged from 0 to 3. Course Plans in Humanities was derived from three sources: (1) the answer to, "If you could, would you take more English courses than the required four years?"; (2) a check on whether or not the student planned to take a foreign language; and (3) a check on whether or not the student wished to take more history or social studies courses than the two years required. One point was scored for each involvement in the humanities, with the scale ranging from 0 to 3.*

Liking for and Grades in Courses. All students were asked to rate a list of courses from 1 to 5 according to how well they thought they would like future courses in that area. They were told to pretend that they didn't know who the teacher would be, only

---

*Unfortunately, the question about English was inadvertently left off the second-year questionnaire and had to be compensated for (in that year only) by adding an extra point where many more humanities courses than required were shown.

that the course was in a particular subject. On this scale 1 represented "I expect to dislike it very much"; 2 represented "I expect to dislike it somewhat"; 3 represented "I am indifferent"; 4 represented "I expect to like it somewhat"; and 5 represented "I expect to like it very much." All students rated biology, chemistry, English, history, languages, mathematics, and physics. Similarly, all students were asked to predict the grade they thought they would receive in each of these subjects, from an F (scored as 1) to an A (scored as 12), including pluses and minuses. Like/Dislike of Humanities was computed as the mean of liking for English, languages, and history or social studies; Expected Grades in Humanities was scored as the mean of predicted grades in the same subjects. Like/Dislike of Mathematics and Expected Grades in Mathematics served as measures of similar constructs for mathematics. Like/Dislike choices were solicited in all three years of testing, and Expected Grades only in the last two years. Because the Expected Grades variable in each area was difficult to interpret (are expected grades indicative of ability or attitude?), it is not discussed in the results.

Career Plans. Students were asked to list their first three choices for a career. Each choice was then rated for the level of education required: 0 represented a high school diploma, 1 represented a college degree, and 2 represented an advanced degree. Each choice was also rated according to subject area — physical science, natural science, or other.

Attitude and Anxiety Scales. Items from four pairs of attitude scales were meshed to form an attitude inventory in the first year of testing, and items from two additional scales were included in the second and third years. In the first year the scales described mathematics and English as easy/difficult, creative/dull, useful/not useful, and enjoyable/anxiety-provoking. In the second and third years additional scales probed the degree of sex-typing of each discipline and perceived encouragement for each discipline from parents, teachers, and peers.

Students were requested to strongly agree, agree, express uncertainty, disagree, or strongly disagree with each statement. A strongly negative feeling toward a subject (whether this was expressed by strong agreement or strong disagreement with a statement) was coded as 1; a somewhat negative feeling ("agree" or "disagree") was coded as 2; indifference ("uncertain") was coded as 3; a somewhat positive feeling ("agree" or "disagree") was coded as 4; and a strongly positive feeling ("strongly agree" or "strongly disagree") was coded as 5. By coding responses in this way, we captured the *sense* of the student's response, regardless of the wording of the item. For example:

Item	Response	Attitude Toward Math	Coding
"Math is difficult"	"Agree"	Somewhat negative	2
"Math is easy"	"Disagree"	Somewhat negative	2

The scores for each item could then be summed to provide a measure of the student's attitude toward mathematics or English.

The first three pairs of scales were primarily derived from those used by Husen (1967) and Fennema and Sherman (1976). One pair of attitude scales (Math is Easy/Difficult, English is Easy/Difficult) contained items concerning the difficulty or accessibility of mathematics and English. Ten items asked about mathematics (five negative, five positive) and eight about English (four negative, four positive). Such items as "I'm not the type to do well in math" and "English is a difficult subject" are typical of these scales.

A second pair of attitude scales (Math is Creative/Dull, English is Creative/Dull) dealt with the degree to which students see mathematics or English as challenging, interesting, and fun, as well as the degree to which they feel they can contribute their own ideas as opposed to memorizing what others already know. These scales used such statements as "Mathematics can be exciting" and "My own ideas cannot be used in English" to elicit students' opinions. Again, there were five negative and five positive items concerning mathematics, and four negative and four positive items concerning English.

A third pair of attitude scales (Math is Useful/Useless, English is Useful/Useless) dealt with the usefulness of the two subjects for everyday life outside of school and for future job and life plans. Such statements as "English is a worthwhile and necessary subject to study" and "I will use math in many ways as an adult" were typical items. Again, ten items concerned mathematics; eight concerned English.

The fourth pair of scales (Math is Enjoyable/Anxiety-Provoking, English is Enjoyable/Anxiety-Provoking) were simplified forms of items on the Mathematics Anxiety Rating Scale (Richardson and Suinn, 1972; Suinn et al., 1972) and on the Fennema and Sherman (1976) Math Anxiety Scale, with parallel items constructed for English. Representative items included "Math doesn't worry me at all" and "It bothers me to have to take English all the way through high school." Twelve statements (six positive, six negative) comprised each scale.

The four pairs of scales just described were used in all three years of testing; two other pairs were used only in the second and third years. One of these examined the sex-typing of mathematics and English, and included the Fennema and Sherman (1976) Math as a Male Domain Scale (renamed Math is Open to All/A Male Domain) as well as a comparable measure for English constructed for this study. The point was to determine the degree to which students felt each subject was more appropriate for one sex to study than the other. Such statements as "It's hard to believe a female could be a genius in mathematics" and "Men are certainly sensitive enough to do well in English" were useful indicators. Twelve mathematics items and 12 English items were included in this pair of scales.

The last pair of scales (Support/No Support from Others—Math and Support/No Support from Others—English) was derived to assess students' perceived support from parents, teachers, and peers. The 12 mathematics and 12 English items adhered to the following patterns: My (parents, teachers, peers) (expect, do not expect) me to do well in (math, English); My (parents, teachers, peers) (encourage, do not encourage) me in my study of (math, English). The phrasing varied somewhat, however, to alleviate boredom.

As we constructed most of the scales specifically for this study, it is important to notice that they are "good" scales: they seem to measure some idea reasonably well. A scale with this property is called reliable, meaning that it doesn't yield greatly different results for very similar populations or for the same population given the test a second time. In a situation like ours, an index called a split-half reliability is used to assess each scale. A high score on this index (close to 1) means that students answer questions on one-half of the scale in a manner similar to the way they answer questions on the other half. Table 2.3 gives the split-half reliabilities for the scales; the reliabilities are very high for all grades, ranging from a low of .59 on Math is Creative/Dull for sixth graders to a high of .92 on Math is Useful/Useless for eleventh graders. Most reliabilities are between .7 and .9, indicating a great deal of consistency in student responses.

Stereotypes v. Self-Image. The last set of questionnaire items, given only to tenth through twelfth graders, asked students to describe their perceptions of a mathematician and a writer and then to describe themselves, using a series of 21 bipolar items first

33

TABLE 2.3
Reliabilities of Attitude Scales, Primary Sample

Scale	6	7	8	Grade 9	10	11	12
Math is Easy/Difficult	.78	.84	.80	.83	.84	.86	.86
Math is Enjoyable/ Anxiety-Provoking	.84	.85	.84	.80	.85	.86	.85
Math is Creative/Dull	.59	.72	.77	.72	.80	.83	.80
Math is Useful/Useless	.85	.81	.81	.86	.85	.92	.87
Support/No Support from Others—Math	—	.74	.78	—	.75	.82	—
Math is Open to All/ A Male Domain	—	.83	.84	—	.83	.89	—
English is Easy/Difficult	.75	.87	.83	.88	.88	.89	.83
English is Enjoyable/ Anxiety-Provoking	.75	.75	.72	.79	.80	.76	.82
English is Creative/Dull	.67	.74	.77	.70	.73	.73	.77
English is Useful/Useless	.79	.78	.83	.76	.82	.85	.78
Support/No Support from Others—English	—	.75	.79	—	.75	.84	—
English is Open to All/ A Female Domain	—	.79	.78	—	.82	.84	—
Number of Students	341	300	275	342	274	239	302

employed by Beardslee and O'Dowd (1961) in their research on members of a large number of professions. Examples of items are:

Sensitive	1 2 3 4 5 6 7	Insensitive
Not at all competitive	1 2 3 4 5 6 7	Competitive
Rational	1 2 3 4 5 6 7	Irrational

In scoring, we determined the distance between the student's self-rating and his or her rating of a writer for each item, and the distance between this same self-rating and the student's rating of a mathematician. Finally, these distances were summed across all 21 items for each occupation. In this way one derives a distance between self and mathematician and between self and writer so that a smaller sum implies a greater identification with a member of that profession.

Essays

We chose to construct essay questions for students to give them an opportunity to tell us what experiences had contributed to their opinions about mathematics and English. We wanted not to repeat the probing of the questionnaire, but rather to expand upon it and come to a better understanding of why students felt very positively or negatively toward these school subjects. The following six questions were asked in the second year of data collection, three each about mathematics and English.

1. Think about your work in mathematics (reading, writing, spelling, and grammar). Think of a lesson or a unit that you really liked. It could be from this year's work or a year in the past. What kind of problems were you working on? (Describe the lesson assigned.) Why did you like it so much?
2. Now think about a lesson or unit in math (English) that you really disliked. What kind of problems were you working on? (Describe the lesson.) What made it so unpleasant?
3. If you were teaching math (English), what is the first thing you would change from the way it is done now? Remember that you want students to enjoy learning math (English) and to be successful at it.

Students were given one-third to one-half of a page to write their answers, and 10 to 15 minutes to think about the best answer.

In order to summarize the answers of students, we read through all of the responses and constructed categories that seemed to capture the threads of students' ideas. These categories will be presented in later chapters, primarily as supporting documentation for numerical results and as anecdotal evidence to confirm a suspicion that arises from the questionnaire results.

Interviews

In the spring of the second and third years of testing we selected the 48 girls whose plans to enroll in mathematics and attitudes toward the field had changed the most from the preceding year so that we could interview them in depth about the reasons for these changes. One-half of each group of 48 showed an increased desire to study mathematics or increased plans for enrollment; the other half showed a dramatic decrease. In each year eight girls were chosen in each grade in each school system, four of whom showed an increased desire for participation, and four a decreased desire. In the last year of testing we also made certain that each group of four girls included two with high ability in mathematics and two with lower ability. This was accomplished by checking their percentile scores on the mathematics achievement test given by their school system and selecting girls who scored as far above or below their school means as possible.

The point of these interviews was to go beyond the information that could be solicited by questionnaire. They covered areas of personal experiences with mathematics that were not discussed in detail on the questionnaire. In the second year of testing we asked students to analyze the differences between their mathematics teachers in the present year and the past year, and to talk about the connection between those differences and their increased liking for or disliking of mathematics. Then we asked them to talk about their parents and siblings. Did their parents give them a lot of encouragement in math? Did their siblings help them with homework? Did their parents expect them to do well in math? Did their parents see them as better or worse than their siblings at math? Finally, we asked about their relations with peers. Who would they ask for help in school? Did they talk with boyfriends about math? Did they like other students who were good in math? Was it all right to be a top math student or was it considered a little strange? Who were the best math students?

In the third year of the study we repeated the questions about teachers because they had elicited the most worthwhile information the year before. We omitted the parental encouragement questions, however, because all students had said their parents were encouraging. Instead, we asked a series of questions about the usefulness of mathematics, trying to find out just where girls felt mathematics would fit into their lives outside of school. We then asked a slightly different set of questions about their peers' reactions to mathematics. We wanted to know, first, the kinds of situations outside the classroom in which mathematics is mentioned and, second, the relative importance for these girls of mathematics and academics as opposed to social concerns.

The interview information, like the essay answers, is used primarily to substantiate the questionnaire results, and certainly gives a broader view of the results than the numbers alone could do.

Standardized Ability Tests

The last source of information on students was their school records. With parental permission we collected their most recent IQ test scores; mathematics and verbal achievement scores; and, for high school students, scores on the Differential Aptitude Tests'

Space Relations section. Since school officials did not wish to have students retested in these areas, we had to accept the scores from instruments that each school system had chosen to use and convert the achievement test scores to appropriate national percentiles. All three systems gave the Differential Aptitude Tests (DAT) to eighth graders late in the school year so we were able, through this one test, to get mathematics and verbal achievement scores, a comprehensive score convertible to an IQ, and space relations scores for almost all high school students. IQ scores for the middle school students came from a wide variety of IQ tests, achievement scores from the Stanford Achievement Tests, or the Iowa Test of Basic Skills (mathematics and verbal scores); no space relations scores were available.

All of this information on ability is useful as an additional descriptor of the students in the sample and in the analysis of the predictors of desire to enroll in mathematics. Ability will be discussed in each of these roles in later chapters with special attention to its effects on participation: Is it the case that only the more talented students plan to take mathematics?

SUMMARY

The research project on which the remainder of this book is based centers on the attitudes and behaviors of 239 sixth graders and 275 ninth graders as they progress through three years of schooling, and on 302 twelfth graders in just that one year of their high school careers. Each student was given an extensive questionnaire in the fall of each year of his or her participation, an instrument which asked about background factors, course plans in a variety of subjects, career plans, and attitudes toward and anxiety about the study of mathematics and English. In the second year of testing, students were also asked a series of essay questions about their work in mathematics and English. In the spring of the second and third years of testing, 48 girls were interviewed to gain more in-depth information about the reasons for their course plans and attitudes toward mathematics. At the end of the three years of testing, then, a large and rich supply of data is available from which we can tease the story of the development of students' attitudes toward mathematics and other school subjects.

3
Students' Attitudes Toward Mathematics

In presenting the results of this study, we are interested in the content of students' attitudes toward mathematics, changes in content across school grades, and differences in attitudes between the sexes. We are also interested in the role of attitudes (along with ability and background factors) in the prediction of students' choices to enroll in optional mathematics courses. In this chapter we will look at the issues surrounding the content of students' attitudes; in Chapter Four we will discuss the socioeconomic status (SES) and ability levels of students and combine this information with the attitude measures to predict participation; and in Chapter Five we will talk more explicitly about the implications of the results for teachers.

In this chapter we will first deal with the cluster of attitudes surrounding Like/Dislike of Mathematics, then turn to the usefulness of mathematics for jobs and everyday lives, and finally examine factors within the social milieu which impinge on the study of mathematics. Within each of these areas we will discuss the positive or negative tone of students' answers on the attitude scales, and then go on to talk about changes in these attitudes across school grades and sex differences in student opinions. Where appropriate, we will add information from the essay questions and interviews to supplement the results from the questionnaires. This discussion of the study results focuses on the primary sample, that is, the twelfth graders and those students for whom we have three years of data. The results for the comparison sample differ only minimally from those of the primary sample and do not refute any of the general conclusions drawn in this or subsequent chapters.

A few introductory comments about the statistics and the results which are presented are appropriate.* In the discussion of grade and sex differences, we have chosen first to present the results on graphs, and then to support the visual evidence with statistics about the level of significance of any differences. One graph is presented for each of the measures used. Figure 3.1A, for example, is a graph of the results for the measure of Like/Dislike of Mathematics (see page 44). On this graph the means (or arithmetic averages) for the boys in the sample are represented by squares (□) and for the girls by filled-in circles (●). Separate means are presented for grades six through twelve, but it should be remembered that the children whose mean is graphed in the sixth grade are the same children whose mean is graphed for the seventh and eighth grades. Similarly, the ninth-, tenth-, and eleventh-grade means are for the same students. Thus, the figures should be interpreted as graphs about three groups of students, two of which have been followed for three years.

The graphs presented for each of the scales used in this study give us a visual image of the differences in student attitudes across grades and between the sexes, and are a valuable first step to understanding the data. The second step is to employ appropriate

---

*The reader who is knowledgeable in statistics may prefer to pass over this discussion.

statistical tests to determine whether the differences we can see are of such a large size that we should worry about them — in other words, to see whether the means of different groups of students differ in a statistically significant manner.

The questions we would like to answer statistically are the following:

1. Do attitudes toward mathematics become more or less positive with the age of the student?
2. Are boys more or less positive about mathematics than girls?
3. Does any age/sex group (for example, ninth-grade girls) stand out from the others as especially negative or positive about mathematics?

To deal with these questions we have used two forms of a technique called analysis of variance, one a cross-sectional test, the other longitudinal.

In general, an analysis of variance examines differences between means. It can be used to evaluate the differences in means for *different* groups of students — for example, the seventh versus the tenth graders, males versus females — or for the *same* students tested repeatedly — for example, the sixth, seventh, and eighth graders. If the set of means being assessed involves different groups of children, the technique is called a cross-sectional analysis of variance. If some of the means involve the same children, the technique is appropriately modified to be a longitudinal or repeated measures analysis of variance. In this study cross-sectional analyses are used to look at males and females in the sixth, ninth, and twelfth grades; in the seventh and tenth grades; and in the eighth and eleventh grades. This kind of test is used for the data generated in each year of the study. Longitudinal analyses are used to look at somewhat different configurations of means — those for males and females in the sixth, seventh, and eighth grades or the ninth, tenth, and eleventh grades. These tests are used to follow a group of children throughout their participation in the study.

Let us take one particular example to illustrate the workings of an analysis of variance, the cross-sectional analysis of Like/Dislike of Mathematics for males and females in the seventh and tenth grades. In order to address the issue of grade differences, we can examine differences between the means for all students in the seventh grade and the means for all students in the tenth grade (regardless of sex). In order to assess sex differences, we can look at the means for all males versus all females (regardless of grade).

These tests of the differences between means for the two grades or the two sexes are called tests of the main effect of the analysis of variance, and yield an F-ratio. The higher this number, the more different are the means under examination. In addition, we can test the "interaction" between the main effects of grade and sex. This means that we can look at the differences among four means: seventh-grade boys, seventh-grade girls, tenth-grade boys, and tenth-grade girls. This test of differences (also yielding an F-ratio) will tell us if any age/sex group stands out from the others because it has an especially high or low mean.

For example, it might be that all seventh graders and the tenth-grade boys like mathematics very much, but that tenth-grade girls hate math. The analysis of seventh versus tenth graders for the main effect of grade might show that tenth graders like mathematics somewhat less than seventh graders; the analysis of the main effect of sex might show that girls like mathematics less than boys; but the analysis of the interaction of sex and grade would clarify that the group of older girls especially dislikes mathematics. Although this example deals with a cross-sectional analysis, its explanation works for longitudinal analyses as well. They too generate F-ratios for the main effects of grade and sex and their interaction, while taking into account the fact that the means for the three grades involved come from the same children.

To answer our three questions thoroughly, we need both the cross-sectional and longitudinal analyses because differences in attitudes among certain grades of children require cross-sectional analyses, among other grades, longitudinal analyses. By looking at the results of these analyses in tandem, we can see how the complete progression from sixth through twelfth grade works. Each of the cross-sectional and longitudinal analyses will also provide us with information on sex differences; each will judge differences between different groups of boys and girls. Similarly, each will judge the interaction of sex and grade to determine if any age/sex group stands out as peculiar.

There is just one more complication in the reporting of these results on attitudes. In the analyses of grade differences when there are three years of data, it is possible for the change in means to follow a straight line or a curved path. If it follows a straight (or linear) path, the means steadily decrease or increase across the three-year period. If it follows a curved (or quadratic) path, however, the means change in one direction from the first year to the second year and in the reverse direction from the second year to the third (making a U-shaped curve or an inverted U). Therefore, in these analyses we must evaluate both the linear change of attitudes across grades and the quadratic change. This method calls for discussion of the following five effects:

The main effect of sex: Are boys different from girls?

The linear effect of grade: Do the means across grades change in a straight line?

The quadratic effect of grade: Do the means for grade go up and then down, or vice versa?

The interaction of sex and the linear grade effect: Does one sex follow a different straight path from the other?

The interaction of sex and the quadratic grade effect; Does one sex follow a different curve from the other?

Now, armed with this introduction to a statistical technique, let us turn to the results.

THE CLUSTER OF PLEASURE/DISPLEASURE WITH MATHEMATICS

Four measures can be included in this cluster of attitudes which differentiate students who like math and seem to get pleasure from its study from those who really dislike math and find little pleasure in its study. The first is the variable called Like/Dislike.* As you will recall from Chapter Two, students responded by choosing a 1 if they thought they would strongly dislike any future math course, a 2 if they thought they would dislike it somewhat, a 3 if they were indifferent, a 4 if they thought they would like it somewhat, and a 5 if they though they would like it a lot.

The other three measures are the attitude scales which were devised to sample opinions about the degree to which students find mathematical material and classes easy as opposed to difficult, enjoyable as opposed to anxiety-provoking, and creative or open to new ideas as opposed to dull or full of memorization. Each of these scales has ten items with which a student can strongly disagree, disagree, express indifference, agree, or strongly agree. In the reporting of the results we have chosen to use item means which have a range from 1 to 5. That is, we added up a student's score on all 10 items of each of these scales and then divided each score by the number of items in the scale. This reporting makes it

---

*All variable names are capitalized to distinguish them as the particular constructs that we have used in the study.

easy to compare the Like/Dislike score with the three attitude scale scores because all four measures share the same numerical range. All one must remember is that a high score from 4 to 5 indicates a positive attitude about mathematics, a score of 3 indicates a neutral position, and a low score of 2 or 1 indicates a negative attitude.

Figures 3.1 and 3.2 are graphs of the results for each of these four attitude measures: 3.1A reports on the Like/Dislike ranking, 3.1B on the Easy/Difficult scale, 3.2A on the Enjoyable/Anxiety-Provoking scale and 3.2B on the Creative/Dull indicator. Table 3.1 shows the number of children whose scores contributed to the means which are displayed and need to be analyzed.

TABLE 3.1

Number of Students in Each Sex–Age Group[a]

| | Grade | | |
Sex	6–8	9–11	12
Male	147	92	143
Female	128	147	159

[a]Only sixth to eleventh graders for whom we have all three years of data are included on the graphs and in the analyses.

The graphs show several important trends in students' attitudes toward mathematics. First, most of the means which appear on these graphs are between 3 and 4. That is, students' general feelings about the study of mathematics are neutral to slightly positive. Sixth to twelfth graders do not express wild enthusiasm or love of the field, but neither are they loath to express good feelings when they have them.

Second, it is clear that students' attitudes toward mathematics become less positive over time. All sixth-grade means are between 3 and 4 on the graphs, but even in the seventh grade, one of the means (girls' scores on the Enjoyable/Anxiety-Provoking scale) is below the neutral point, and more means drop below 3 over the years.

Third, boys are consistently more positive about mathematics than girls when the issue is Easy/Difficult and Enjoyable/Anxiety-Provoking. Boys are quite clear in their opinions that mathematics is fairly easy and enjoyable, not difficult nor very anxiety-provoking.

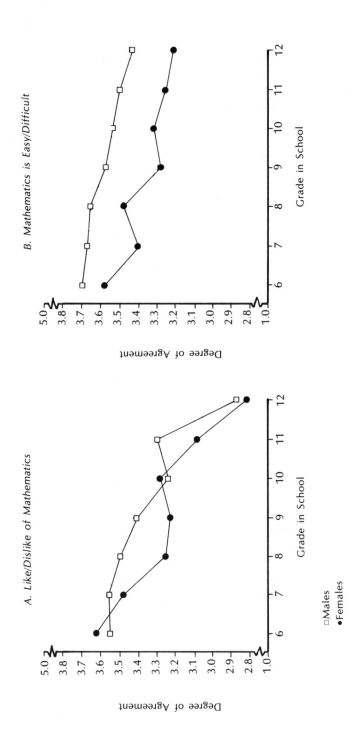

FIGURE 3.1.

The Attitude Factors of Like/Dislike and Easy/Difficult

FIGURE 3.2.

The Attitude Factors of Enjoyable/Anxiety-Provoking and Creative/Dull

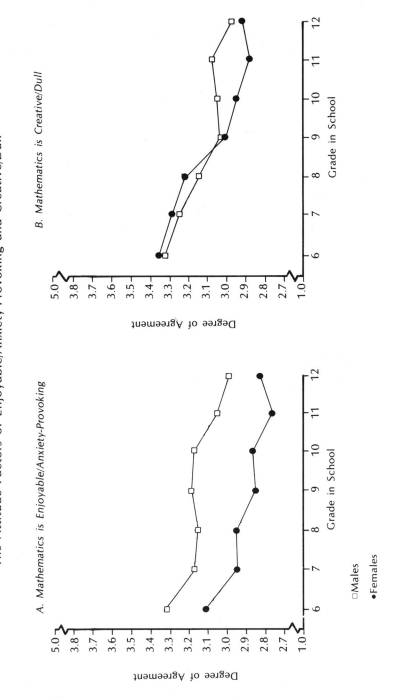

A. Mathematics is *Enjoyable/Anxiety-Provoking*

B. Mathematics is *Creative/Dull*

□Males
●Females

We looked at separate graphs depicting students' attitudes toward English and found that they do not follow the same patterns as Figures 3.1 and 3.2. The lines depicting attitudes toward English are comparatively flat (Like/Dislike, Easy/Difficult), showing no change over time, or they slope upward (Enjoyable/Anxiety-Provoking, Creative/Dull), showing a general increase in positive attitudes across grades. There is something about the study of mathematics which makes it less pleasant as time goes on, but attitudes toward English are maintained or improve. Thus, changes in attitudes toward mathematics are not reflections of changes in attitudes toward all school subjects, but rather are peculiar to mathematics (and possibly some other disciplines).

The statistical analyses summarized in Tables 3.2 and 3.3 substantiate the trends which we observed on the graphs: the cross-sectional and longitudinal analyses of variance resulted in several consistent grade and sex findings. First, if we look for consistency

TABLE 3.2

F-Ratios for Main Effects and Interactions for Cross-Sectional
Analyses of Mathematics Attitude Scales

| | Grade Comparisons | | | | | | | | |
| | 6/9/12 | | | 7/10 | | | 8/11 | | |
Attitude Scale	Sex	Grade	S × G	Sex	Grade	S × G	Sex	Grade	S × G
Like/Dislike	.07	9.55***	.46	.04	6.28**	.52	3.50	2.07	.04
Easy/Difficult	18.21***	14.71***	1.17	15.90***	2.98	.05	13.67***	9.62**	.14
Enjoyable/Anxiety-Provoking	18.05***	12.61***	1.47	20.38***	.69	.27	16.48***	4.45*	.52
Creative/Dull	3.00	28.86***	2.18	.27	18.66***	1.11	.84	12.72***	3.81
Useful/Useless	8.06**	13.37***	2.10	3.92*	5.99*	.05	1.61	19.16***	2.21
Support/No Support from Others	NA			7.13**	.32	.08	.73	10.40**	.07
Math is Open to All/A Male Domain	NA			59.57***	1.54	.06	82.28***	.27	.16

*p ≤ .05
**p ≤ .02
***p ≤ .001

in grade differences, one of these four attitude scales (Creative/Dull) shows regular differences. In Table 3.2 there are highly significant grade differences in all three cross-sectional comparisons, and in Table 3.3 the pattern is continued with two significant linear grade effects on the longitudinal comparisons. Thus the following statement can be made:

There is a continual and significant decline in students' attitudes toward the nature of mathematical work. As they grow older, students feel more and more that their role in mathematics is to memorize what other people tell them rather than to think things through on their own and contribute their own ideas. Concomitantly, mathematics seems less interesting, challenging, and fun the further students get in its study.

TABLE 3.3

F-Ratios for Main Effects and Interactions for Longitudinal Analyses

Attitude Scale	Sex	Grade (linear)	Grade (curved)	Sex x Grade_L	Sex x Grade_C
Sixth–Eighth Grade:					
Like/Dislike	.53	3.93*	.30	1.81	.02
Easy/Difficult	7.85**	1.76	2.86	.66	2.69
Enjoyable/Anxiety-Provoking	9.14**	13.28***	3.59	.01	.29
Creative/Dull	.36	11.31***	.04	.08	.08
Useful/Useless	1.47	.15	.00	2.23	.36
Seventh–Eighth Grade:					
Support/No Support from Others	2.65	.64	NA	1.59	NA
Math is Open to All/A Male Domain	38.85***	5.45*	NA	1.02	NA
Ninth–Eleventh Grade:					
Like/Dislike	.46	1.32	.00	.00	4.24*
Easy/Difficult	12.69***	.62	1.58	.49	.53
Enjoyable/Anxiety-Provoking	18.72***	6.49*	2.40	.16	.02
Creative/Dull	5.06*	7.98**	.43	.01	2.18
Useful/Useless	7.39**	17.61***	.76	.99	1.46
Tenth–Eleventh Grade:					
Support/No Support from Others	1.32	7.76**	NA	1.26	NA
Math is Open to All/A Male Domain	51.24***	.02	NA	1.78	NA
Distance of Self/Mathematician	.46	11.40***	NA	.15	NA

*$p \leq .05$
**$p \leq .01$
***$p \leq .001$

Second, in each of the other instances of a significant grade effect, the direction of the effect is consistent: students like mathematics less and less as they grow older, and students find mathematics more difficult and more anxiety-provoking with increasing age. So, although the findings are not completely consistent for grade differences in the Like/Dislike, Easy/Difficult, and Enjoyable/Anxiety-Provoking scales, the trends are clear, and the following statements can be made:

There is a decrease in students' liking for mathematics with age on the straightforward index of degree of like or dislike for the field.

With increasing age students are more likely to say that mathematics tends to be difficult and that they may not be able to do advanced work in the field.

As time goes on, more and more students admit to being uncomfortable and even anxious in quantitative situations. More agree that tests are unpleasant, and that it is problematic to find enjoyable lessons in mathematics.

Next, if we look for consistently significant sex differences, the tables show that there are large and regular differences in all of the sex difference analyses for math being Easy/Difficult and Enjoyable/Anxiety-Provoking. The following can be stated about the results of each of the five analyses:

Girls claim that mathematics is more difficult than boys do and girls rate themselves as more anxious in quantitative situations.

On the other hand, sex differences do not appear in any of the five analyses of Like/Dislike and appear in only one of the analyses of Creative/Dull. Boys and girls rate their general level of liking for mathematics similarly, and they have similar feelings about how interesting the study of mathematics can be.

Last, in looking at the interactions between sex and grade, we can see that only one (the interaction of sex and the curved grade effect of Like/Dislike for ninth to eleventh graders) is significant, and even that one is not terribly strong. Thus the differences between means in these analyses are predominantly due to the main effects of grade and sex that have already been discussed, rather than to some particular age/sex group of students who are different from everyone else.

The most important conclusions to be reached from all four graphs and the accompanying statistical tests, then, are the following:

Grade differences occur in all four of these measures of attitudes; older students are more negative about mathematics than younger students.

Sex differences are large and consistent for two math scales, Easy/Difficult and Enjoyable/Anxiety-Provoking; girls are more negative about mathematics than boys.

In attempting to define just what it is about mathematics that bothers students, we can consult their responses to the essay questions on the second-year questionnaire and the interviews in years two and three. Students' answers to the essay questions about favorite topics and their reasons for liking/disliking lessons are summarized in Tables 3.4 to 3.6. In Table 3.4 we have clustered the reasons why students say they like or dislike math, and have reported the percent of students who spontaneously expressed each reason. In Table 3.5 we have listed the topics students liked or disliked the most, and in Table 3.6 the ways they would change mathematics teaching if they could. In addition to listing topics and changes, we have reported the percentage of students who mentioned each topic or change.

TABLE 3.4
Reasons Students Give for Liking or Disliking Mathematics[a]

Reasons for Liking Math	Percent of Students Giving This Reason
Easy/Understand material/Do well[b]	41
Fun/Challenging/Interesting[b]	36
"Just like it"	7
Useful material	4
Good teacher	4

Total number of students giving at least one reason: 903

Reasons for Disliking Math	
Difficult/Don't understand material/ Do poorly[b]	53
Time-consuming/Boring/Too much memorization[b]	22
"Just dislike it"	8
Poor teacher	6
Useless	4

Total number of students giving at least one reason: 878

[a]Answers from students in both the primary and comparison samples are included in this table.
[b]These are three ways of expressing the same idea. Some students used all three, some used only one. We classified their reasons by meaning rather than state the specific words used.

Students in all schools expressed common sentiments about mathematics classes. They primarily disliked topics when the material was hard or difficult to understand and they did poorly, and they liked topics which were easy to understand, on which they

TABLE 3.5

Topics in Mathematics that Students Especially Like or Dislike[a]

Topics Liked	Percent of Students Naming This Topic
Fractions/Decimals/Percents	21
Algebra/Word problems	16
Addition	13
Geometry	11
Multiplication	8
Division	7

Total number of students naming at least one topic: 1,001

Topics Disliked	
Fractions/Decimals/Percents	24
Division	14
Algebra/Word problems	11
Proofs	9
Geometry	6

Total number of students naming at least one topic: 916

[a]Answers from students in both the primary and comparison samples are included in this table.

could do well. As one seventh grader wrote, "When I really understand things, I can like math better than English." This comment may seem a bit negative toward mathematics, but it does illustrate that students believe it is important not to feel mathematics is too difficult, but rather to feel competent in and comfortable with the material.

Other frequently cited reasons for liking mathematics were that the topics and classes were fun, challenging, and interesting, and other reasons for disliking lessons were that they were time-consuming or boring, or that they involved too much memorization. So these students enjoy being presented with problems which make them think, they derive satisfaction from successfully answering a tough question, and they do not enjoy repetitive assignments which require them to solve a series of very similar problems. Although these reasons seem simple, they may pose a bit of a dilemma for teachers who would like to respond to them,

TABLE 3.6

Students' Suggestions for Changing the Teaching of Mathematics[a]

Suggested Change	*Percent of Students Making This Suggestion*
Explain material thoroughly/ Make simpler	19
Make it more fun, exciting, and interesting/Get more student input	14
Use games and puzzles	10
Assign less homework and more in-class work	9
Be nicer/Offer more personal help outside of class/Be more encouraging	8
Allow students to work at own pace	6
Relate material to "real life"/ Teach importance of mathematics	4
Total number of students making at least one suggestion: 734	

[a]Answers from students in both the primary and comparison samples are included in this table.

because there is a fine distinction between material which is challenging and that which is too difficult, as well as between material which is easily understood and that which is boring.

Students' concerns with the easiness of the material, their competence in it, and the pleasure of doing an assignment are also reflected in their choices of favorite topics in mathematics (see Table 3.5). Even in high school, some basic arithmetic operations—addition or the manipulation of fractions, decimals, and percents—were commonly mentioned as favorites. High school students had mixed feelings about algebra and geometry—both come up on the lists of topics most liked and most disliked—and showed a clear dislike for working out proofs, whatever the theorem to be judged. All too many students had to search far back in their memories to find math topics that they could honestly say they enjoyed, and tapped memories of arithmetic in elementary school to find some lessons that were understandable and fun.

When students were asked how they would change the way courses are now taught (see Table 3.6), a predominant theme emerged: Make sure all students understand the material. Over and over again, they stressed that teachers should have many explanations for how to do a problem and why an operation works the way it does. They said that if they were teaching, they would make sure that all students' questions were addressed during class and that everyone felt free to seek help from teachers during and after class. They suggested that a class format where students could interrupt to ask questions was really necessary so that any confusion could be cleared up before further material was presented.

They also said that classes without the usual lecture format were especially pleasant. Some said that if they were teaching they would make mathematics class more fun by adding games, puzzles, and projects. They liked interesting approaches to problem solving, especially those which allowed them to assume an active role in the learning process and express their own ideas. Though the precise activities which would promote such a role were not defined, the flavor was clear: Give us material we can understand in a style that facilitates learning by being responsive to our own ideas, interests, questions, and needs.

These answers to the essay questions do give some insight into the reasons for the decline in positive attitudes toward mathematics over time. Classes are getting more formal, and the use of puzzles and projects has become rare. The material being introduced is getting more difficult, and teachers may just not be as accessible to high school students as they are to the younger groups. The more typical lecture format is not as conducive to free question-asking, so students may not be clearing up initial confusions by obtaining clarification from teachers. They may prefer living with the confusion to interrupting the progress of a class with a question which they think may be considered stupid. This, of course, would serve to make the material still harder—and not at all fun.

The formality of the classroom also produces some anxiety for students. When there is not an easy interchange between teachers and students, teachers are forced to rely on homework assignments and tests as indicators of the degree of understanding students possess. Some students reported that their discomfort with homework resulted because they didn't understand what they were supposed to do, and most students expressed dislike for frequent tests. Some claimed to "freeze" on mathematics tests; many were frustrated by an inability to finish, no matter how well they had prepared. In suggesting how they would alleviate these anxieties, students did not often propose the abolition of homework and tests; they seemed to recognize the value of formal evaluative methods. The students did, however, offer suggestions for reducing the pressures created by these situations. Some mentioned that it was helpful for them to keep notebooks of mathematical procedures to which they could refer when doing homework problems or studying for tests, and said they would require these in their classrooms. Others said that if they were teachers, they would assign homework problems early enough in the class period so that there would be time to explain the assignment and respond to questions. These students want to understand mathematics and want to like it, but find it hard to do either in formal classrooms.

It seems, then, from students' essays and interview comments, that the reasons for the downward trend in attitudes across grades is due somewhat to the increasing difficulty of the material, and also to the increased formality of the mathematics classroom, where students feel uncomfortable about asking too many questions.

The reasons for sex differences are not as clear from the interviews and essays, although a couple of possibilities were suggested. It may be that girls are more conscious of the image they present in class, and less likely than boys to ask a question which might be "stupid." This, of course, would result in their finding the material more difficult and in their being more anxious in class, knowing that they may have to speak up and not wanting to present the wrong image.

Another reason for the sex differences may be differences in the answering patterns of boys and girls on the questionnaire items themselves. Some of the items on the Easy/Difficult and Enjoyable/Anxiety-Provoking scales leave themselves wide open to bravado. For example, on the Easy/Difficult scale, two key items are "I am sure that I can learn mathematics" and "I can get good grades in mathematics." The Enjoyable/Anxiety-Provoking scale includes such items as "I usually have been at ease during math tests," "Math doesn't worry me at all," and "It wouldn't bother me at all to take more math courses." At each grade level the males score more positively than females on these items. (All scale items and the means for students of each sex at each grade level are presented in Appendix B, Tables 1–6). It could be that they really do find math easier and more enjoyable, but the sex difference could very well be exacerbated by the boys' tendency to downplay any inability or stressful reaction to mathematics. There is, in fact, some evidence that, regardless of the subject matter, boys rate themselves as experiencing less anxiety than girls rate themselves, even in instances when trained clinicians can discern no differences between the sexes in the anxiety they feel (Maccoby and Jacklin, 1974; Sarason and Winkle, 1966).

However, our data do not support the general conclusion that boys always rate themselves as less anxious than girls rate themselves. Specifically, the data from the measure of English as Enjoyable/Anxiety-Provoking suggest that, compared to girls, boys rate themselves as experiencing as much or more anxiety in English. It may still be the case, of course, that some particular bias built into the subject matter of mathematics encourages boys to deny anxiety, but we cannot support the notion that this trend is universal.

While the "real" sex differences in attitudes toward mathematics may be reduced in degree from the large differences in means reported in this section, we must acknowledge the fact that differences probably exist and that these may contribute to girls' lowered desire to participate in optional mathematics courses. Both sexes become more negative about mathematics with age, and girls are particularly sure that mathematics tends to be difficult and produce anxiety. In Chapter Four this problem will be discussed thoroughly in the analysis of the attitudinal contributors to students' desire to take more mathematics courses.

THE PERCEIVED USEFULNESS OF MATHEMATICS

The next idea that we need to investigate is the degree to which students see mathematics as something which can be useful to their lives outside of school and to their search for jobs in the future. As Figure 3.3 demonstrates, students from sixth to twelfth grade look upon mathematics as having a great deal of potential usefulness. The means range from over 3.5 to slightly over 4.0, clearly in the "useful" area of the graph, above a neutral score of 3.0. An examination of the means for the individual items (see Appendix B, Table 4) shows that the highest degree of agreement occurs on the general item "Mathematics is a worthwhile and necessary subject" and on the job-related item "Knowing mathematics will help me earn a living." Strong disagreement is regularly voiced to the idea that "Taking mathematics is a waste of time." Thus, students are looking at mathematics as a practical tool for the job market, and see the subject as generally useful knowledge to possess.

FIGURE 3.3.

The Pragmatic Factor of Usefulness

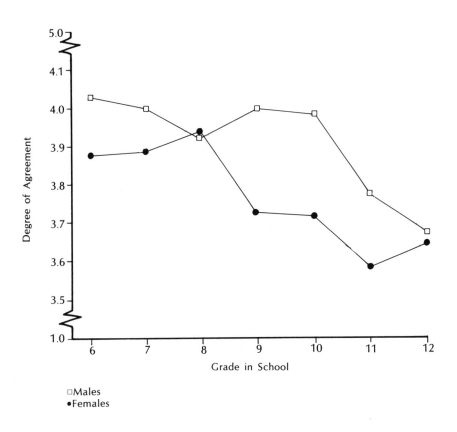

As with attitudes toward mathematics, there are grade and sex differences in the degree to which students support this notion of usefulness, though at all ages girls and boys see mathematics as useful. A look at the graph suggests that the notion of usefulness declines with age (particularly through high school), and that there are some sex differences, with boys more favorable than girls. These sex differences, too, seem strongest in high school. These observations are supported by the consistently significant grade differences in the cross-sectional analyses summarized in Table 3.2 and in the longitudinal analysis of high school students shown in Table 3.3. The sex differences are not as substantial as was the case, for example, with Easy/Difficult, but they appear in two of the cross-sectional analyses and in the longitudinal analysis of high school students. Hence, we can conclude that the degree to which students think mathematics is useful decreases with age, particularly in high school, and that high school girls see mathematics as less useful than do high school boys.

To explore students' notions of the usefulness of mathematics a bit further, we can look at the results from the essays and interviews. Four percent of the students who wrote essays giving reasons for liking mathematics cited usefulness as a criterion (see Table 3.4) and another 4 percent cited lack of usefulness as a reason for disliking the subject. These students all expressed concern over the roles the subjects they were learning would play in their lives. They liked learning things that they could see a later use for, and disliked topics that seemed to lead nowhere. In discussing how they would change the teaching of mathematics (see Table 3.6), 4 percent of the writers specifically suggested relating the information to "real life" situations and explicitly detailing the importance of the current topic. Although the percent of students discussing usefulness was not large, it indicates a vocal minority.

When the girls interviewed in the third year were explicitly asked about the usefulness of mathematics, they had very strong opinions. About half of them felt that knowledge and skills in mathematics would be valuable for their career aspirations, either in the practice of a career or in the training for it. A few of these students wanted to go into mathematical careers, such as computer work or mathematics education, and saw mathematical training as central to their career goals. More commonly, though, students were considering some area of business, for example, accounting or business management, and felt that some knowledge of mathematics would be important in those careers. Other students said that they wanted to go into medical careers, such as veterinary medicine, pharmacy, or nursing, and that mathematics would be helpful preparation for the course work required for these careers. One tenth-grade girl interviewed was planning to be a flight attendant, and said she was thinking about writing to an airline school to see what background in high school mathematics was necessary or recommended. She said that the confirmation that mathematics would be serving her future needs was the most convincing reason she had to take more courses.

The girls who were interviewed were almost unanimous in their feeling that mathematics would be useful for some aspect of their everyday lives. When asked to mention these aspects, they suggested figuring out family budgets, balancing checkbooks, filling out income tax forms, and doing the weekly shopping. A few said that mathematics could also be helpful in cooking, building things, or sewing.

It is interesting to note that these examples require only the skills of *arithmetic;* none require more advanced knowledge of mathematics. When the high school students were asked specifically about *higher mathematics,* most said that it was not necessary to their planned careers or everyday lives. Addition and subtraction would be needed, but the algebra and calculus they were currently learning were not going to be useful later on. And the students who said that mathematics would be useful to their careers were unable to specify ways in which it might be helpful. They could only suggest that "math is always a part of adult life," or "math is very functional." So middle and high school students see arithmetic as useful for their planned careers and later lives, but even the older students do not feel that knowledge of and skills in more advanced areas of mathematics could be relevant to their adult pursuits.

One might choose to argue that students' assessments of (advanced) mathematics as not being useful to the future is essentially accurate. Students may be choosing to go into fields in which mathematics is not used to any great extent, or may be choosing not to go to college but rather into jobs which require only a high school diploma. Female students in particular may be choosing not to pursue careers, feeling that it is more important to work in the home and raise a family. The information which was gathered on career aspirations of students can help us examine this possibility. All career choices were analyzed for required level of education and required degree of mathematical knowledge so that we could investigate the level of mathematics that would legitimately be needed.

Each student's preferred career was coded as requiring (1) a high school education, (2) a college degree, or (3) graduate work. The *Occupational Outlook Handbook's* classification of the usual amount of education attained by people in these careers was used as a guide. Figure 3.4A presents the level of education required for the careers to which students aspired. Since the results across grades were similar, we have combined all grades for this presentation. As can be seen, the plurality of male and female students chose careers which require only a high school diploma, such as being a secretary, carpenter, or professional athlete. A fair proportion of students chose careers which require a college degree, and girls were more likely to choose such careers than boys. A relatively small proportion of students in all grades chose careers which require a graduate degree; boys were slightly more likely to aspire to such careers than girls, especially in the later years of high school. Thus, the majority of students were not thinking about the prospects of becoming Ph.D.s, but needed high school preparation for immediate entry into the job market.

Figure 3.4B illustrates the degree to which students desired careers requiring knowledge of and skills in mathematics. To create this table, each career was categorized as not openly mathematical (nonscience), requiring some mathematics in preparation for the career (natural sciences), and necessitating some mathematical training for work which would be quantitative in nature (physical sciences). Each category contained jobs at all levels of required education, so the coding of area of intended career was fairly independent of amount of education. For example, the physical sciences category included the carpentry and electrical trades, college-level careers in computer programming and civil engineering, and advanced-level careers in astronomy and mathematics. In general, Figure 3.4B shows that a plurality of students chose careers not related to science, including the small percentage of students (both males and females) who indicated that they preferred to be a husband or wife and maintain the home rather than pursue a career. Among the remaining students, girls were more likely to aspire to careers related to the natural sciences (for example, beautician, nurse, oceanographer) than careers related to the physical sciences, and boys were more likely to choose careers related to the physical sciences (for example, carpenter, engineer, astronomer) than those related to the natural sciences.

A look at the relationship between the level of education required for a career and the degree to which mathematics is used in it helps to explain the reasons for the existence of sex differences in career aspirations. That is, a check on the amount of education required for the boys' careers in the physical sciences shows that their preference for this category is primarily due to their interests in entering the trades, those physical science careers requiring only a high school diploma. Similarly, the preference of girls for careers in the natural sciences is due to their aspirations to careers requiring only a high school education—beautician or animal shelter worker.

If we look beyond careers requiring a high school diploma, the sex differences in preference for mathematically related careers disappear. Boys aspire somewhat more frequently than girls to college-level careers related to mathematics, for example, engineer

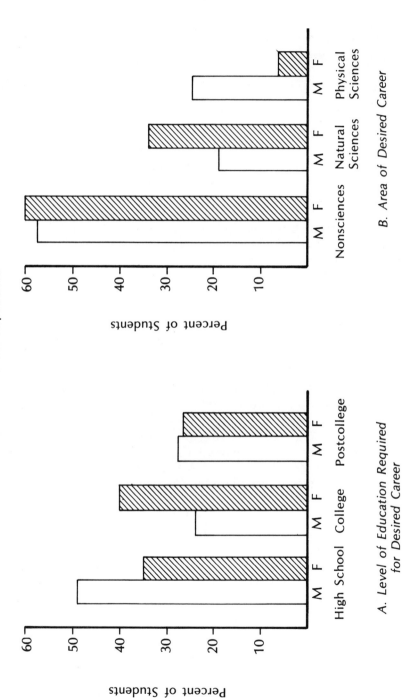

FIGURE 3.4.
Students' Career Aspirations

and computer scientist, but this difference is not statistically significant, and there is little sign of a sex difference in the postcollege physical science careers—only a handful of students of either sex aspire to those.

Thus, the picture which seems to emerge is that throughout middle and high school, about half of the students would like to enter careers which require no education beyond high school. In keeping with traditional career patterns, boys are more likely to desire jobs related in some way to mathematics, such as carpentry, electrical work, and auto mechanics. Girls, on the other hand, are more likely to aspire to jobs which do not relate to mathematics, whether they be nonscience careers, such as flight attendant or secretary, or natural science careers, such as beautician or animal shelter worker. A relatively small percentage of students wish to pursue professional careers related to mathematics, and this appears to be equally true of boys and girls.

Looking at these career findings, one might argue that students' assessments that mathematics will not be useful to their future careers are actually fairly astute, particularly for girls. First, one could argue that many students do not plan to go into jobs which require more than a high school diploma, so they do not need four years of high school mathematics as preparation for mathematics courses in college or graduate school. Second, most students do not plan to enter careers which are strongly related to mathematics. This is particularly true for girls, of whom a maximum of only 10 percent in each grade planned to enter careers strongly related to mathematics, implying that mathematics—beyond arithmetic— probably won't be needed in their future lives. Third, most students who do plan careers which require some quantitative knowledge and skills are planning to enter the trades, and these careers will not require much knowledge beyond arithmetic.

If, then, we take our jobs as educators to be preparing students for the place they wish to assume in society, we may be reduced to asking why we try to teach anything beyond addition, subtraction, multiplication, and division. Can any parts of algebra, geometry, trigonometry, or calculus be useful to these students? The answer is certainly yes, but the reasons for this answer must be explored carefully, and the arguments presented above soundly defeated.

First, we must accept that knowledge of mathematics is more than just the ability to perform simple calculations by following a set of rules. It also implies an understanding of the operations performed. Algebra and geometry (and, one could argue, all of higher mathematics) give a deeper appreciation of why arithmetic operations work. For example, most of us have seen children add up a column of figures and arrive at an absurdly small sum. They may have an adequate knowledge of the rules for lining up a column of figures, for summing the numbers in each column, and for "carrying" numbers to the next column, but their knowledge seems to end with these mechanical rules. They do not seem to know why the rules work, or to understand such principles as the sum must be larger than any of the numbers being added. The study of inequalities in algebra (the reasons certain sums must be bigger than others and why) and the work with representations of wholes in geometry can — if all goes well — give students that extra insight into why and how mathematical operations work. It is this insight and the thorough understanding of operations it implies which will insure the correct solution of mathematical problems.

Second, in a more philosophical vein, mathematics teaches the ability to solve problems, to reason from premises to a conclusion. This skill is necessary in all aspects of adult life. The process of proving a theorem involving two parallel lines or simplifying an algebraic equation necessitates a clear view of the problem which must be solved and the facts at hand which may help to solve the problem. It also requires that a student work from the facts he or she has in a logical, step-by-step progression to a defined end point. Facility with such a pattern of logical argument is useful in a wide range of adult situations, from convincing a boss of the necessity of purchasing new equipment to persuading a spouse that one's point of view is the only rational perspective on the world.

Third, some of the techniques and ideas taught in high school mathematics *are* directly applicable outside of the classroom. For example, dressmaking and carpentry are much simpler for students who understand the symmetries of figures and the principles about angles that are taught in geometry. Carpenters benefit from knowing that if they cut a piece of trim in order to achieve a 40° angle, the piece that remains will have a 140° angle. Shortcuts for figuring out complicated inventory needs can make use of algebraic equations where x's and y's indicate interlocking parts. Perhaps one needs to order enough parts to repair 30 washing machines, and each machine needs 1 pump and 23 screws. There are 10 pumps and 116 screws in stock. How many should be ordered?

Some algebraic facts may even be directly useful. Suppose a pharmacist is told that a prescription for poison ivy requires that a patient take eight pills the first day, seven the second, and so on to the last pill. The pharmacist who remembers high school algebra will know that the sum of a series of consecutive integers is equal to n times $n + 1$, divided by 2, where n is the highest integer in the series. Students need to be made aware of the uses for the mathematical ideas and techniques they are learning. Applications do exist, and students planning careers that do not—on the face of it—demand mathematics may find that their high school study has direct applications.

Fourth, we want to expand students' job opportunities, not limit them. Many of the students in this study will enter the trades or become beauticians, but as stated earlier, that doesn't mean they will continue in those occupations. A beautician may shift to office work and be expected to do the bookkeeping or manage the inventory on a small computer. The auto mechanic may decide that a white-collar job is attractive and become a computer technician or programmer. Such career changes will be much more difficult for students who have discontinued their study of mathematics as soon as they could than for those who were offered the opportunity of taking the business math and algebra which would put them at ease in the more mathematically oriented positions they may later wish to enter. Helping students to understand a subject and encouraging them to keep options open may also help them realize that mathematical knowledge can be useful. Teachers are in an excellent position to show them just how this is so.

INFLUENCES FROM THE SOCIAL MILIEU

Finally, we must consider the social influences which may impinge on students in their study of mathematics. Three measures of these influences were employed in the study: the degree of support students thought they were receiving from teachers, parents, and peers; the degree to which students felt that mathematics was a male domain; and the distance students felt existed between their own personalities and that of a mathematician. Data from these measures are less extensive than those from the earlier measures; the first two measures were included only on the second- and third-year questionnaires, and the last was given only to the oldest group of students in each year's testing. All of the available information is displayed in Figures 3.5 and 3.6 and in Tables 3.2 and 3.3.

As with the earlier variables, the means of students' reports of perceived support from others (see Figure 3.5A) fall in the neutral to positive range. Both males and females in the seventh to eleventh grades feel that their parents, teachers, and peers encourage them in mathematics. From this graph it seems as though some change in the amount of support may have occurred during stu-

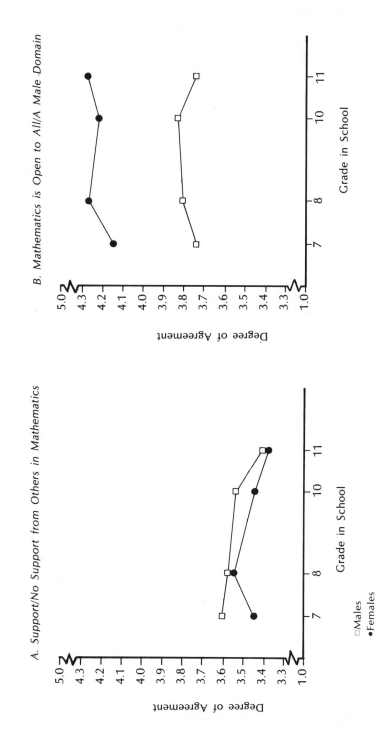

FIGURE 3.5.

The Social Factors of Encouragement from Others and the Sex-Typing of Mathematics

FIGURE 3.6.

Distance Between Self-Concept and Stereotype of a Mathematician

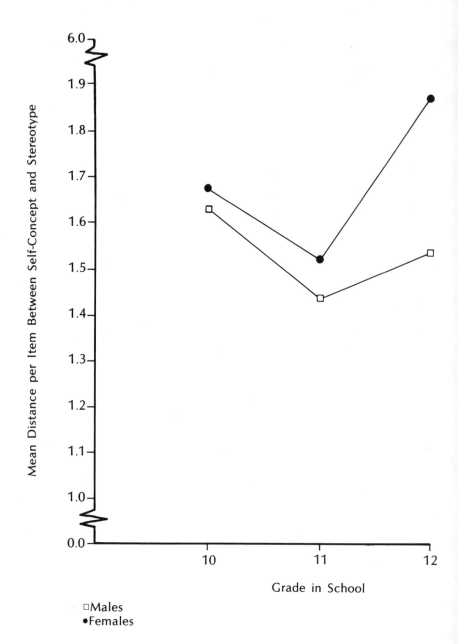

dents' high school years but little in the middle school years, and that sex differences in perceived support are minimal. These observations are substantiated in Tables 3.2 and 3.3. There were significant grade differences in the cross-sectional analysis of eighth versus eleventh graders and in the longitudinal analysis of tenth and eleventh graders, suggesting a decrease in support across the high school years. And there was only one sex difference — in the cross-sectional analysis of the seventh- and tenth-grade scores. In this case females felt they had less support than males felt they had. Thus, there is little evidence from this measure of the social milieu that boys and girls are encouraged differentially in mathematics, and little evidence that encouragement differs for the grades below eleventh. This does not mean that the sexes actually receive identical treatment (to make that conclusion we would have had to observe teachers, parents, and peers interacting with those students), but it does imply that boys and girls do not believe that the significant people in their environment are treating them very differently.

The graph describing whether students consider math to be open to all or a male domain appears on Figure 3.5B. The means plotted on this graph indicate that all students think mathematics is appropriate for both sexes; the means range from a low of about 3.7 to a high of over 4.2, representing a very open-minded stance. The tests of significance summarized in Tables 3.2 and 3.3 show that there are no grade differences in thoughts about the appropriateness of mathematics for both sexes, as is apparent on the graph, but there are huge and consistent sex differences. Girls are always more certain that mathematics is appropriate for everyone.

The reasons for the lack of grade differences and the strong sex differences may be that this scale is tapping an opinion which all of our society knows is required — you are supposed to say that women and men are equally competent in all fields, whatever your age and actual beliefs are, and no matter what your behavior would seem to imply. Girls, in particular, may be responding to the questions on this scale with what they feel are appropriate "liberated" answers. When they are asked whether they would "trust a woman just as much as . . . a man to figure out important calculations," they immediately answer in the way which has become socially desirable, perhaps without thinking whom they elected class treasurer or which parent in their household makes out tax returns. Very few females can agree with the statement "It's hard to believe a female could be a genius in mathematics" or "Girls who enjoy studying math are a bit peculiar."

It may also be that these students firmly believe that mathematics *should* be appropriate for both sexes and that women *can* be as good at mathematics as men. These are, after all, the phrasings of the questions on the scale, and students did strongly agree with the scale items. At the same time, they may recognize that fewer women than men are known as geniuses in mathematics and fewer and fewer women will be enrolled in mathematics classes as they get older. In other words, they may simultaneously hold this very positive belief about the possibilities of study in mathematics, but still recognize that the current situation is not as positive. Girls may conclude from these conflicting messages that they *could*, of course, choose to study more mathematics, but they won't. Thus, this measure may not be tapping a central reason for participating in or avoiding mathematics, but only a belief peripheral to the issue. A more central measure might be a request for students to observe the number of students in advanced mathematics courses and in mathematically oriented careers and then to rate the degree to which this knowledge influences their own course plans.

The third measure of the effects of the social milieu which we used was the distance score established between each student's self-image and his or her stereotype of a mathematician. The results for this index are presented in Figure 3.6 and in Table 3.3. The numbers indicating distance on the graph are the mean distance a student feels from a mathematician, derived from his or her total distance score divided by 21, the number of items on the scale. The potential range of distances is 0 to 6 and the smaller the distance, the more positive is the student's attitude toward mathematics. Thus, the means on the graph imply that students feel they share some but not many qualities with mathematicians.

Only two statistical analyses of means could be conducted from these data. The first is a longitudinal analysis of tenth and eleventh graders which shows no sex differences, but a closing of the gap between self-image and stereotype of a mathematician with increasing age (see Table 3.3). The second is a test of the difference between boys and girls in the twelfth grade which shows boys to be significantly closer in self-image to a mathematician than girls [t(298) = -4.63, p < .001]*. Unfortunately, the results here

---

*A t-statistic is the square root of an F-ratio, and is used when only one main effect is under examination and only two groups are being compared. The farther a t-value is from 0 in a negative or positive direction, the more significant is the difference between the means being compared. This one is relatively far from 0, indicating a substantial difference between boys and girls.

are inconsistent for both the grade and sex effects. The longitudinal analysis of grade showed a significant decrease in distance, but the twelfth-grade means on the graph reverse the trend. The longitudinal analysis of sex differences showed no strong difference between girls and boys in the tenth and eleventh grades, but the analysis of twelfth graders showed a difference.

These social factors do not follow the pattern of systematic grade and sex differences that appeared for the other attitude variables. There does seem to be a decrease in the amount of support students feel they receive during high school, but grade trends in the other measures of the social milieu were not consistent. The only consistent sex differences were in answer to the question of math being open to all or a male domain, where females were *more* positive about mathematics rather than less, in contrast to previous scales. These factors in the social milieu of students do not seem to differentiate those in different grades or of different sexes.

One explanation of the lack of differences may be that we haven't asked the best questions, a suggestion we made when discussing the results for Math is Open to All/A Male Domain. Another may be that the social environment actually influences everyone the same way. But to accept this explanation we need much more evidence, which should come from careful observations of the people in students' environments and their interactions with the students. A third explanation may be that our method of asking students to describe the forces around them is simply inappropriate; social influences operate at such a subconscious level that no one can "truthfully" answer direct questions about their effects.

Not only were our attempts to examine features of the milieu unsuccessful on the questionnaire scales, but our essay and interview questions also turned up little information of interest. Social influences were never spontaneously mentioned in the essays students wrote. In the interviews, girls gave no evidence of receiving messages different from those boys received. Asked who they usually went to for help on math problems, most girls answered with the name of a girlfriend or their teacher, but they all said that they wouldn't feel funny about asking a boy in their class the questions—even the smartest boy, if he was a nice person. When asked if they would like to date a boy who was good in math, they said it really didn't matter how he did in math, that what he was like as a person was more important. When asked if students who are good in math are strange, the girls said they are no more strange than students who are smart in general. We looked for differences between those students who had steady boyfriends and those who didn't, and found that there weren't patterns of increased participation in mathematics among either group, but rather that dating was the norm in some schools and not in others.* We also attempted to find groups of girls who differed on their relative weighting of the importance of having a career and having a family, thinking that the two groups might be able to talk about different experiences. We could not, however, find a "career group" of any significant size. When all 48 of the girls in the third-year interviews were asked if they would rate career as more important than family, equally important, or less important, 33 said family came first, 11 said the two were equally important, and only 3 rated career goals as more important than family. The high value placed on family characterized the girls' interviews as a whole, regardless of mathematical ability.

Social influences are simply difficult to measure. The approach we adopted—asking students what sorts of things impinge on them—did not elicit many differences. Measuring influences of the environment through observations may be more successful, but unfortunately this was beyond the scope of this study. We will still look at the relationship of these measures to participation, but with a question in our minds about the thoroughness of our measurement of such influences.

SUMMARY

In sum, the information gathered through students' interviews, essays, and questionnaires over the course of a three-year period seems to indicate the following attitudes toward mathematics:

Students feel that mathematics becomes more difficult with time. They like to feel competent in the field and so prefer topics which they feel are easy to understand and teachers who

*One rural middle school had virtually no girlfriend/boyfriend pairs and there was an almost unanimous feeling among the girls that mathematics was a terrific subject. But we couldn't say that these positive attitudes were attributable to their not dating, because this was the only instance of its type. It was also the only purely rural school, the only school with all female mathematics teachers, and the only school with a mathematics program which had students making contracts with teachers for the work they intended to accomplish and then working on their own to complete the contract. Any one of these variations, as well as their interactions, might be the "cause" of the positive attitudes toward mathematics.

allow them to ask questions freely. Although all students feel that math is reasonably accessible, this attitude is significantly less prevalent among females than males.

Though students do not like mathematics material which is too difficult, they do like material which is challenging and receive satisfaction from solving tough problems. They think math material becomes less interesting and fun as they progress through middle school and high school, and suggest that they, as teachers, would try to infuse students with the notion that mathematics can continue to be fun and challenging, and that it can require creative thought.

Accompanying the increase in perceived difficulty of mathematics are students' increased feelings of anxiety about mathematics as they progress through school, and this is particularly the case among females. Anxiety seems to result more from the evaluative situations frequent in mathematics classes than simply from work with mathematical material, and may be reduced by relaxing the atmosphere of the classroom.

Students think highly of topics and materials which are useful and personally relevant. They see many uses of basic arithmetic skills for their future careers and everyday life, but can see little use for higher mathematics. Uses do exist, but students, even those in high school, seem ignorant of them.

Most students feel they receive a fair amount of encouragement from teachers, parents, and peers in their study of mathematics. As students progress through high school they feel less encouraged but never to the point that they feel they are being discouraged.

All students think of mathematics as an appropriate domain for both males and females, though girls are even more adamant than boys in this view. It remains to be seen whether such a view has any relationship to course participation in mathematics or is simply a reflection of a "liberated" sample of students.

Students seem to feel reasonably close to mathematicians in self-image, recognizing a rational and stable side of themselves. We must wait to find out if the more rational among the students are also the ones who take more mathematics.

This chapter has provided an introduction to the attitude scale results from the study. Such results alone may be useful to teachers as descriptors of students' feelings about mathematics and their classrooms. In the next chapter we ask how these opinions about mathematics influence participation in the field, a critical issue for teachers, guidance counselors, parents, and educators planning intervention programs. How can we use the knowledge of attitudes to influence course plans in mathematics?

πΣπΣπΣ **4** πΣπΣπΣ

Predicting
Participation

In this chapter we report the results of our attempt to define the roles of our variables in predicting students' participation in mathematics, paying special attention to the different roles of predictors for males and females. Earlier chapters have given us a good feeling for students' scores on the potential predictors and on the measure of participation. In this chapter we look more closely at the participation variable, discuss the ability scores which have not yet been presented, review the roles that the background and attitude variables may be expected to play, and then present the findings on the prediction of participation.

COURSE PLANS

Figure 4.1 is a graph of the participation variable, which we have chosen to name Course Plans. Recall that this measure differs for the younger and older students in the sample. The younger students had to circle the subject they would choose to take in a series of pairs of subjects, supposing that they had to take one of the two subjects and couldn't take both. The graph for sixth through

FIGURE 4.1.

Course Plans in Mathematics

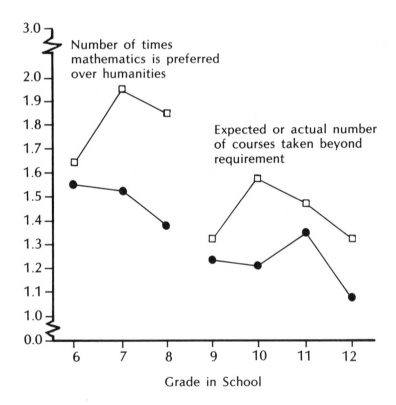

□Males
•Females

eighth graders displays the number of times they said they would prefer to take mathematics over English, social studies, or a foreign language; the range in possible scores was 0 to 3.* The older students wrote down the number of years in high school that they intended to take (or actually had taken) mathematics. For these students, the Course Plans variable was scored as the number of years of intended participation beyond the two years that each high school required. Its range is therefore also 0 to 3, as students could take two years of mathematics in any one year of high school or could accelerate through summer school courses.

The means for the younger students indicate that most would prefer to take mathematics courses over humanities courses about half of the time (that is, most scored over 1.5); boys are particularly favorable toward mathematics. The means for the high school students are also over 1.0, suggesting that many students plan to take at least one year of optional high school mathematics and some are going even further. Again, boys seem to be choosing more mathematics than girls.

Cross-sectional and longitudinal analyses of variance were done on Course Plans to parallel those done on the attitude scales (see Table 4.1). They showed that the sex differences which could be observed on the graphs are significant—boys have a stronger preference for mathematics in all of the analyses. In addition, the cross-sectional analyses indicated that younger students were regularly more willing to participate in mathematics than older students. It is difficult to know how much of this difference is due to the difference in the measures used, so this result must be taken as merely suggestive. The longitudinal analyses demonstrated no significant changes within the sixth- to eighth-grade group or the ninth- to eleventh-grade group, so that the cross-sectional changes may not represent a true decline in students' desire to participate, but simply the difference in indices.

We can look into the relationship of the two Course Plans measures by considering a specific result from the third-year testing. In that year we asked both eighth and eleventh graders to an-

---

*In discussing this variable we refer to a comparison of mathematics and *humanities* because social studies and a foreign language are a part of the measure, as well as English. References to the attitude scale measures are concerned with mathematics and *English* since the questions on these scales only ask about English.

TABLE 4.1

F-Ratios for Main Effects and Interactions
from the Analyses of Course Plans

A. Cross-Sectional Analyses

	Effects		
Grade Comparison	Sex	Grade	$S \times G$
6/9/12	5.43*	6.29**	.62
7/10	17.48***	12.38***	.10
8/11	12.31***	4.85*	3.59

B. Longitudinal Analyses

	Effects				
Grade Comparison	Sex	Grade (Linear)	Grade (Curved)	$S \times G_L$	$S \times G_C$
6/8	11.19***	.03	3.74	4.55	.08
9/11	9.43**	3.38	.06	3.20	.74

*$p \leq .05$
**$p \leq .01$
***$p \leq .001$

swer the preference question *and* the question about number of years of high school mathematics. The strength of the association between the indices can be seen through a statistic called a correlation. Correlations range between −1 and +1, and the further the correlation (or r) is from 0, the stronger the relationship between the indices. An r near +1 means that the indices are measuring nearly identical constructs; an r near −1 implies that the indices are measuring opposites. The actual correlations of the indices of preference for mathematics and number of planned courses are .30 (p<.01*) for eighth graders and .27 (p<.05) for eleventh graders.[1] This implies that the indices are related to each other significantly, and can probably be considered to represent aspects of the same construct, but they are not perfectly interchangeable.

---

*This p-value means that the probability of getting a correlation this large by chance alone is less than 1 in 100; the probability of getting this correlation because there *is* a relationship is more than 99 in 100.

The problem, of course, is that we have little choice in defining the Course Plans measures. Asking sixth graders how many years of high school mathematics they want to take would not have been productive because their answers could not have been interpreted. Children of that age do not know the substance of high school mathematics, the requirements that the high school will place on their studies, or the number of courses that a college of their choice will look favorably upon, so their answers to a question about number of courses would not mean the same thing as the answers of high school students. On the other hand, it did not make sense to reduce the question for the older students to one of preference when we really wanted an index of actual participation. So we are forced to make do with the present measure, recognizing that it is by no means a perfect answer to the problem of how to define participation.[2] Since the primary analyses of why students choose many or few courses are done separately for the two age groups, the principal consequence of the difference in definitions of Course Plans is that the results for the two age groups need to be interpreted somewhat independently.

THE PREDICTORS

Background Variables

Gender. This is a critical variable in the prediction of Course Plans because one primary research question is how gender is involved in the decision to take or avoid mathematics. There are a number of possibilities. It could be that attitudinal and background variables totally account for differences in Course Plans, and that the gender of the student adds no interesting information to the prediction equation. Or it could be that attitudes do not explain the differences in participation, and that being female or male, on its own, seems to contribute to differential participation. Another possible answer is that gender has a role in addition to the other predictors. Each of these possibilities is explored through the testing of prediction equations which include all of the background and attitudinal variables in addition to gender in predicting Course Plans. We then go on to predict Course Plans separately for males and females. This will allow us to see if any of the attitudinal or background factors operate differently for the two sexes.

Socioeconomic status (SES). As we noted in Chapter Two, the SES measure in the study is derived from a coding of the occupation of the head of the student's household.[3] A higher number indicates a higher social class. In this sample nearly half of the students (46 percent) are from working class backgrounds, one-fifth (21 percent) are from white-collar households where parents hold jobs in the lesser professions or in small or medium businesses, and one-third (33 percent) are from households where parents are in the major professions or senior business management positions. The SES of a student's family may act as a predictor of Course Plans because different SES groups may have different expectations of potential jobs and possible degrees. Children in high-SES families may be expected to take mathematics through high school because they will need it for college, so that SES on its own or in conjunction with other variables may operate as a predictor of participation.

Ability

Four measures of ability were solicited from student records for this study: IQ; national percentile on a test of mathematics achievement (Mathematics Percentile); national percentile on a test of verbal achievement—mostly comprehension (Verbal Percentile); and national percentile on the Differential Aptitude Test (DAT), Space Relations section (Space Percentile). The first three scores were collected for all children from tests they took just prior to or at the beginning of the first year of our testing; the last score was collected only for the high school students from their eighth-grade testing. Table 4.2 summarizes the scores.

The mean IQ scores in Table 4.2 indicate that the children in this sample are somewhat above the mean IQ of 100, as defined by the makers of IQ tests. An analysis of variance of these scores showed that they differ for these two grades (F = 10.09 df = 1,499,

TABLE 4.2
Means on Ability Measures[a]

Measure	GRADE 6		GRADE 9	
	M	F	M	F
IQ	105.21	105.23	108.89	109.22
Mathematics Percentile	48.03	41.49	65.19	65.25
Verbal Percentile	56.92	55.27	63.49	61.63
Space Percentile	NA	NA	66.43	64.40

[a]The grades given are six and nine because students were in these grades in the first year of testing when scores were collected. The table means are for students who were tested in all three years of the study.

p<.01), though the absolute size of the difference is small, and they do not differ between the sexes. The grade difference is probably due to the fact that poorer students tend to transfer from the public high school to local trade or vocational high schools, drop out of school, or in other ways make themselves unavailable for three consecutive years of testing.

The Mathematics Percentile scores are considerably above the mean percentile of 50 for the ninth graders, suggesting that these school districts are succeeding at teaching mathematical concepts and computations to their students. The fairly low percentile scores for the sixth graders and the consequently strong age difference in scores (F = 66.95, df = 1,499, p<.001) is due to one school district which chose to wait to teach students fractions and decimals until they were in the sixth grade, but gave an achievement test to their beginning sixth graders which extensively tested them on fractions. The students scored poorly on this exam, but such scoring was not an accurate reflection of their abilities — although it was a reflection of achievement. The report from the guidance offices at the schools which were involved is that the seventh-grade achievement scores were much higher. Unfortunately, we could only collect scores in one year, so we must keep their limitations in mind. In fact, the verbal achievement scores (reading comprehension and vocabulary) are probably a better reflection of the achievement levels of students in these systems, showing all ages to be above the norm of 50. Sex differences were not significant.

The percentiles recorded for the ninth graders on the DAT suggest that these students are very facile with mentally folding up two-dimensional diagrams into three-dimensional figures. The means are considerably above national norms, and sex differences are small.

In deciding to use these measures in prediction equations, we are faced with a small dilemma. All three measures of importance to us (IQ, Mathematics Percentile, Spatial Percentile) tap aspects of the same idea — ability. If we put a number of such similar variables into a prediction equation, they will vie with each other for supremacy in predicting Course Plans, resulting in ambiguities, if not errors, in the analysis. One variable may look as though it is the sole significant predictor when, in fact, a whole set are equally good predictors, and the one that shines through is "suppressing" the effects of the others. To resolve this dilemma we have chosen to combine into one Ability score the IQ and Mathematics Percentile scores for the younger students and all three scores for the older students. This has been done by first standardizing each score and then adding them together. (Standardizing means changing the metric of each measure so that zero is the mean for students, and their scores are spread out around that mean in a particular way.) Standardizing scores gives equal weight to each score. For example, adding a raw IQ score of 106 and a raw Mathematics Percentile score of 58 would give more weight to the IQ score than it should have. The new Ability score can be entered into prediction equations to see if higher-ability students are the ones who want to take a great deal of mathematics.

Attitudes

Eight attitude measures were employed in this study: Like/Dislike, Easy/Difficult, Enjoyable/Anxiety-Provoking, Creative/Dull, Useful/Useless, Support/No Support from Others, Open to All/A Male Domain, Distance Between Self and Mathematician. We would like to know how all of these relate to participation in mathematics, but to use all of them in prediction equations would raise the same problems discussed in relation to Ability. Their content is just too similar to trust the outcomes from the testing of such equations.*

Our solution to the problem of overlap in this case was to eliminate the Like/Dislike measure since this attitude is represented in the other indices of pleasure/displeasure, and then to take a very close look at the relationship between the items on the remaining seven scales through a technique called factor analysis. This technique looks at the way in which students have answered the individual questions on the attitude scales and groups together those items which are answered similarly.[4]

In our case, the first step was to factor analyze the items from all of the mathematics attitude scales in each year of testing. The results of these analyses were that two scales could remain intact, Useful/Useless and Open to All/A Male Domain, since the items on each scale clustered together, but each set was independent of all other sets. In addition, the teacher items from the Support/No Support from Others scale appeared as an independent unit, and could enter into our prediction equation as such.[5] But the Easy/Difficult, Enjoyable/Anxiety-Provoking, and Creative/Dull scales overlapped, as might be expected.

---

*Statistics evaluating the degree of similarity among these measures — correlations — are presented in Appendix B, Table 7.

The specific groupings of items from the overlapping scales to emerge from the factor analysis are identified in Table 4.3. The table displays the factor loadings for each of the items in a group or subscale for each of the grades tested in the study. Factor loadings are similar to correlations in that a higher positive score means a stronger relationship between the item and the subscale to which it belongs. If a factor loading is larger than .40, we have given it an asterisk, meaning that in this particular age group the starred item

TABLE 4.3
Factor Loadings for Items in the Feeling Subscales

	Grade in School						
Items	*6*	*7*	*8*	*9*	*10*	*11*	*12*
EASY							
Math is fairly easy.	0.41*	0.54*	0.37	0.52*	0.62*	0.53*	0.55*
I can get good grades in mathematics.	0.27	0.53*	0.62*	0.64*	0.46*	0.47*	0.69*
Math has been my worst subject.	0.30	0.57*	0.39	0.57*	0.51*	0.46*	0.64*
Math tends to be difficult.	0.30	0.64*	0.30	0.38	0.38	0.26	0.34
I'm not the type to do well in math.	0.48*	0.63*	0.62*	0.61*	0.44*	0.37	0.78*
FEAR							
I get scared when I open my math book and see a page full of problems.	0.49*	0.51*	0.40*	0.40*	0.54*	0.57*	0.44*
I feel nervous during math tests.	0.76*	0.59*	0.44*	0.79*	0.78*	0.71*	0.26
I get a sinking feeling when I think of trying hard math problems.	0.48*	0.60*	0.49*	0.43*	0.54*	0.60*	0.52*
I feel uneasy when I realize I must take a certain number of math classes to graduate from high school.	0.41*	0.59*	0.51*	0.16	0.24	0.32	0.59*
I usually have been at ease during math tests.	0.63*	0.48*	0.46*	0.68*	0.81*	0.78*	0.23
Math doesn't worry me at all.	0.59*	0.64*	0.37	0.49*	0.72*	0.71*	0.43*
FUN							
Calculating percentages is fun.	0.23	0.48*	0.51*	0.44*	0.57*	0.61*	0.35
When a math problem arises that I can't immediately solve, I stick with it until I have the solution.	0.20	0.36	0.25	0.04	0.42*	0.50*	0.10
Mathematics can be exciting.	0.57*	0.53*	0.59*	0.54*	0.64*	0.62*	0.65*
I am challenged by math problems I can't understand immediately.	0.09	0.37	0.28	0.30	0.31	0.48*	0.26
It wouldn't bother me to take more math courses.	0.40*	0.48*	0.57*	0.16	0.41*	0.56*	0.60*
My own ideas can be used in math.	0.45*	0.46*	0.46*	0.32	0.47*	0.49*	0.22
I enjoy being given a set of addition problems to solve.	0.35	0.42*	0.35	0.50*	0.46*	0.39	0.14
I usually enjoy preparing for a math test.	0.43*	0.53*	0.51*	0.41*	0.57*	0.63*	0.28
Math gives me the chance to think things out for myself.	0.40*	0.53*	0.43*	0.35	0.40*	0.52*	0.31
There is not much room for my own ideas in math.	0.28	0.29	0.45*	0.34	0.42*	0.38	0.04
The challenge of math problems does not appeal to me.	0.15	0.29	0.51*	0.39	0.52*	0.54*	0.62*
I put off studying for a math test as long as I can.	0.49*	0.25	0.41*	0.30	0.54*	0.48*	0.17
Math is dull and monotonous.	0.54*	0.47*	0.59*	0.51*	0.62*	0.58*	0.62*
I do as little work in math as possible.	0.42*	0.38	0.40*	0.24	0.44*	0.47*	0.18

*Factor loading greater than 0.40.

clearly belongs on the subscale. Not all items have stars for each age group, but each item within a subscale does have asterisks for at least one age group.

The first subscale, EASY, consists of the five strongest items from the original Easy/Difficult scale. Two items refer to mathematics as a subject and three refer to the respondent's evaluation of his or her own performance in mathematics. The second group, FEAR, is a subset of the original Enjoyable/Anxiety-Provoking scale. Almost all items refer to the respondent's performance on mathematics tasks, two of these referring to tests. The last scale, FUN, contains items from several scales, all of which concern enjoyment in doing mathematical tasks, feelings of involvement and accomplishment, or feelings concerning the attraction of mathematics. To compute students' scores on each of these subscales, we summed their scores on the items within each subscale. Then we checked to see that these were "good" or reliable scales. All three came out well, with reliabilities of over .80 (see Chapter Two for an explanation of reliabilities), which means that they are not quite as reliable as standardized tests, but are very reliable for attitude scales.

The second step in the specification of an independent set of attitude scales was to examine the relationships among these three new subscales to see if they could all be used as predictors or if they needed to be combined in some way (see Table 8 in Appendix B for these correlations). The result of the examination was that the subscales overlap sufficiently to necessitate combining them. Since this overlap made sense substantively—students who feel capable of handling tasks related to mathematics should enjoy their involvement in these tasks and not feel undue anxiety—we combined the three subscales into one variable, Feelings, by summing students' scores on all 25 appropriate items, and dividing by 25. This has given us a strong index of pleasure/displeasure with mathematics, better than any individual scale score (including Like/Dislike), but representing them all.

The final step in preparation for the prediction analyses was to check that the new Feelings variable was reasonably uncorrelated with the other attitude indices that would be included in the prediction equations. The correlations appear in Table 9 of Appendix B, and are sufficiently close to 0 for us to proceed with the prediction analysis. Thus, scores from the attitude scales of Feelings, Useful/Useless, Open to All/A Male Domain, and Support/No Support from Teachers can be added to the background measures of Gender and SES, the Ability measure, and, for the tenth and eleventh graders, Distance Between Self and Mathematician, to predict Course Plans.

DEALING WITH LONGITUDINAL DATA

One additional statistical issue needs to be resolved before we can turn to the results. This is the problem created by a longitudinal study which involves asking students to answer the same questions year after year. While such a method is optimal for looking at changes in students' ideas, it creates the same sort of difficulty that we encountered when working with the ability and attitude measures. That is, if we put the three Feelings variables (one from each year of testing) in the equation as predictors, we would be in trouble because of their content overlap. Similarly, we would have to make a decision about which index of Course Plans we wanted to predict. We could evaluate a separate equation for each year of the study using only the measures given to students in that year, but this strategy would limit our ability to talk about *change* in Course Plans and would not use all of the available information efficiently.

In order to use the maximum amount of data that we can (both years of data for measures of the social milieu, all three years for the other attitude variables and Course Plans) we have chosen to compute a new series of scores for each measure given in more than one year.[6] The first score represents the *level* of a variable, the second represents *linear change* in that variable over time, and the third represents *curvilinear change* over time. These ideas were introduced in Chapter Three and can be expanded here through an example presented in Figure 4.2.

Suppose that Figure 4.2A is a graph of a student's scores for Course Plans over the three years of the study. Figures 4.2B through D illustrate how this pattern would be redefined into its level, linear change, and curvilinear or quadratic change components. The level of this student's Course Plans — represented by the horizontal line in Figure 4.2B — is the average or mean of the scores that the student received on Course Plans for the three years of the study. The linear change component in Figure 4.2C is also a straight line, but slopes upward, representing the difference in the student's Course Plans for the first and third years of testing. This student planned to take more mathematics courses in the third year. The quadratic or curvilinear change score is represented in Figure 4.2D.

FIGURE 4.2.
Dividing Longitudinal Data into Component Parts

A. Theoretical Results

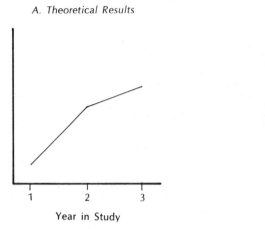

Year in Study

B. Level Component[a]

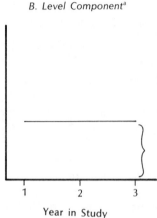

Year in Study

C. Linear Change Component[a]

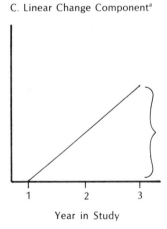

Year in Study

D. Quadratic Change Component[a]

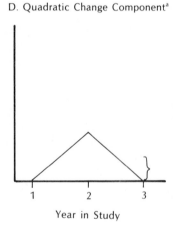

Year in Study

[a]The brackets show the part of the graph that is measured by each component.

Its measure is based on the deviation of the pattern of scores from a straight line. Its shape in this example is due to the fact that the student changed more from the first year to the second year than from the second year to the third. The degree of increase in Course

Plans seen from the first year to the second was not maintained from the second year to the third. Rather, the increasing trend leveled off.

For each of the variables for which we have three years of data (Course Plans, Feelings, Useful/Useless) we have defined separate variables to represent level, linear change, and quadratic change scores. For each variable with two years of data (Support/No Support from Teachers, Math as Open to All/A Male Domain, Distance Between Self and Mathematician) we have computed level and linear change scores only. (Three pieces of data are needed to define quadratic change.) We are now prepared to define a set of prediction equations using this refined set of measures to predict level and change in Course Plans over the course of this longitudinal study.

RESULTS FROM PREDICTING COURSE PLANS

We have organized the analysis of Course Plans by the age of the students, so first we will discuss the findings for the middle school students and then those for the high school students. Within each age group we will discuss the following series of analyses:

1. Analyses of the level of and linear change in Course Plans for both sexes from the second to the third year of the study. Such analyses are needed because we only have these two years of data on the social factors, and we want to know how they, along with the other potential predictors, influence Course Plans. To include them in any analysis, we must limit other variables to the same two years of information.

2. Separate analyses of the data from year two to year three for males and females. These analyses will allow us to specify the different roles which our predictors might play for the sexes.

3. Analyses of the level of and the linear and quadratic changes in Course Plans from year one to year three of the study. All predictors for which we have three years of data plus the background variables can be included in these equations.

4. Separate analyses of the data from all three years of testing for males and females.

Each of these analyses uses what is called a regression approach.[7] Such a technique tries to apportion the variance in the outcome of Course Plans into parts which are attributable to the different predictors. In our case we want to know whether the variance in Course Plans between male and female students is due primarily to differences in Feelings, to attitudes about Usefulness, to a feature of the social milieu, to Ability, to SES, to Gender, or to some combination of these variables. A regression analysis produces numbers which can inform us about the relevance of each measure in the prediction of Course Plans, as well as the relative importance of one measure as opposed to the others.

Because we are particularly interested in the role of Gender in predicting Course Plans, we have first chosen to look at the role of Gender in the prediction equation of all students, and then to examine differences between separate prediction equations for males and females. The initial combined-sex analyses can tell us if Gender is at all responsible for differences in students' Course Plans or if the variance in Course Plans can be accounted for by differences in such predictors as Feelings or Usefulness without Gender. A significant effect for Gender would imply that boys and girls who are equal in attitudes, ability, and SES still wish to take different numbers of mathematics courses, and would be a serious finding to explain. The second separate-sex analyses then allow us to see if some of the predictors are relevant for only one sex. It may be that Ability, for example, is a strong predictor for one sex, but not for the other. Such an effect may not be significant in the combined-sex analyses, but would appear in the separate-sex analyses. Its appearance—or other variables—would give us more insight into how the sexes differ and how each group decides on Course Plans.

Tables 4.4 through 4.7 summarize the results.[8] Each of the numbers in the tables is the coefficient by which we must multiply a student's score on that variable to raise his or her score on Course Plans by one point. Coefficients that were small enough to be statistically insignificant were omitted from the table to make it simpler to find the very significant ones. The numbers in the row labeled R^2 indicate the amount of variance in Course Plans that is accounted for by the total prediction equation. The R^2 of .42, for example, means that 42 percent of the variance in Course Plans for that group is accounted for by the measures in that analysis. Ideally, we would like an R^2 of 1, as that would mean our predictors totally explained Course Plans, but R^2's over .2 are acceptable, and values over .4 are good.

Middle School Students

Two-year analyses. First, let us look at the two-year combined-sex analyses of Course Plans for the seventh and eighth graders (see Table 4.4). These will provide the backdrop against which other

TABLE 4.4

Prediction Results Using Two Years of Data

	Grade 7/Grade 8		Grade 10/Grade 11	
	Level of Course Plans	Linear Change in Plans	Level of Course Plans	Linear Change in Plans
Level of Feelings	.29***			
Change in Feelings		.32***		
Level of Usefulness				
Change in Usefulness				
Level of Teacher Support				
Change in Teacher Support				
Level of Math is Male Domain				
Change in Math is Male Domain				
Level of Self/Mathematician	NA	NA		
Change in Self/Mathematician	NA	NA		
Ability			.30***	
Gender	.28***		.34**	
SES			.10***	
R^2	.42	.26	.45	.07
Number of Students in Analysis	271	271	187	187

**$p < .01$
***$p \leq .001$

findings can be compared. The level of Course Plans is explained primarily by the level of Feelings one has about mathematics and by one's Gender. This suggests the following conclusions:

The first major predictor of the level of students' Course Plans or, in this case, choice of many courses in mathematics over courses in the humanities, is the student's general level of Feelings about mathematics. Students who are more positive about the field more frequently prefer mathematics courses to humanities courses.

A student's Gender is the other major predictor of level of Course Plans. Males in this younger group have a stronger preference for mathematics courses over humanities courses than do females, even after we have accounted for differences in Feelings, Usefulness, Teacher Support, Math as a Male Domain, Ability, and SES. The number on the table implies that middle school boys who are equivalent to their female class-mates on all other predictors score .28 higher on Course Plans than their classmates. Because the Course Plans variable only ranges from 0 to 3, this difference of slightly more than one-quarter of a course is relatively large.

None of the other predictors in this equation account for a large portion of variance in Course Plans. Usefulness, Teacher Support, Math as a Male Domain, Ability, and SES are not major predictors of level of Course Plans for middle school stu-dents. In a positive vein, then, we can say that relatively low-ability students and students from working-class homes have not ruled out mathematics as a subject to study.

To extend these findings, we can look at the prediction of level of Course Plans for the seventh/eighth graders separately for males and females (see Table 4.5). As was true in the analysis of all stu-dents together, level of Feelings is a major predictor, and Useful-ness, Teacher Support, and Math as a Male Domain do not enter as significant predictors. But in these equations, SES is a significant predictor for boys and Ability for girls, implying that different fac-tors affect participation for the two sexes.

In addition to having more positive Feelings about mathemat-ics, boys who frequently choose mathematics over humanities courses tend to come from families with *lower* socioeconomic status. We can conclude, then, that a lower SES does not act to preclude the study of mathematics for boys. These middle school students see a good future for themselves in the study of this field and would like to take many courses.

In addition to having more positive Feelings about mathe-matics, girls who frequently choose mathematics over human-ities courses tend to score *lower* on Ability. They, as well as the boys from lower SES families, have not typed themselves as failures and see no reason why they shouldn't go on to partici-pate in more mathematics courses.

The next analyses to review are the predictions of change in Course Plans for the seventh/eighth graders (see Table 4.4). With the sexes together in the equation, only one predictor is significant: changes in Feelings toward mathematics. The more a child's plea-

TABLE 4.5
Separate Prediction Results for Males and Females Using Two Years of Data

	Grade 7/Grade 8				Grade 10/Grade 11			
	Level of Course Plans		Linear Change in Plans		Level of Course Plans		Linear Change in Plans	
	M	F	M	F	M	F	M	F
Level of Feelings	.30***	.29***						
Change in Feelings			.38***	.28***				
Level of Usefulness								
Change in Usefulness				.06***				
Level of Teacher Support								
Change in Teacher Support								
Level of Math is Male Domain								
Change in Math is Male Domain								
Level of Self/Mathematician								
Change in Self/Mathematician								
Ability		−.28**			.28**	.31***		
SES	−.14***					.12***		
R²	.44	.43	.32	.31	.53	.41	.08	.10
Number of Students in Analysis	143	128	143	128	65	122	65	122

**p ≤ .01
***p ≤ .0001

sure with mathematics increases, the more that child changes in preference for taking mathematics courses. This result indicates that if educators want to try to change students' thoughts about taking mathematics, they can aim at changing students' feelings about the difficulty of mathematics or the degree to which it is fun and creative, and expect that changes in preferences for courses will follow. The results suggest that even though boys and girls differ in their general level of preference for mathematics courses, their preferences change in the same way over time. So the same method of influencing one sex to change Course Plans should work with the other sex.[9]

The separate-sex analyses of linear change in Course Plans over this two-year period (see Table 4.5) second these notions about intervention. They show again that change in Feelings is the primary predictor of changes in Course Plans and, consequently, that intervention programs might aim to change these attitudes. But, in addition to this variable, the regression for girls demonstrates that a change in attitudes about the Usefulness of mathematics predicts a change in Course Plans. This finding can be translated into another recommendation:

> The educator who would like to plan an intervention program for girls to change their Course Plans, might try to alter their notions of how useful mathematics can be, because this study suggests that girls whose ideas of the usefulness of mathematics blossom tend to want to take more mathematics courses.

Three-year Analyses. By and large, the three-year analyses support these conclusions from the two-year analyses (see Tables 4.6 and 4.7). Rather than reiterate these findings, we will discuss the

TABLE 4.6

Prediction Results Using Three Years of Data

	Grade 6/Grade 7/Grade 8			Grade 9/Grade 10/Grade 11		
	Level of Course Plans	Linear Change in Plans	Curved Change in Plans	Level of Course Plans	Linear Change in Plans	Curved Change in Plans
Level of Feelings	26**					
Linear Feelings		.26***				
Curve of Feelings			.34***			
Level of Usefulness						
Linear Usefulness						
Curve of Usefulness						
Ability				.33***		
Gender						
SES				.10***		
R^2	.42	.27	.30	.41	.06	.06
Number of Students in Analysis	271	272[a]	271	187	187	187

**$p \leq .01$
***$p \leq .001$
[a]One child for whom we lacked year two data on Course Plans was not included in the level or curve analyses, but was included in the linear change analysis.

places where the two sets of analyses differ. This occurs in the combined-sex analysis of the role of Gender in predicting level of Course Plans and in the single-sex analysis of the role of Ability in predicting girls' level of Course Plans. The following are the specific differences and probable explanations for their existence:

Gender was a significant predictor of level of Course Plans in the two-year analyses, but was not significant when the sixth-grade scores were included. The reason for this may be that Gender has no predictive role in the sixth grade, but only becomes important in the seventh grade. When the sixth-grade scores are added to the data set, they mitigate the role of Gender for all three grades. Seventh grade could thus be a critical time for intervening to stop males and females from taking different paths with regard to mathematics.

Ability did not enter for girls in predicting level of Course Plans in the three-year analyses as it did in the two-year, but the interpretation of this finding is the same as it was then: lower-ability girls are still considering taking a lot of mathematics. They have not ruled it out of their curriculum.

The three-year analyses also extend the two-year findings in that they contain analyses of the curvilinear or quadratic changes in Course Plans in addition to the linear change analyses. A check on these results (see Tables 4.6 and 4.7) provides us with two further conclusions:

The quadratic change in Feelings is a strong predictor of the quadratic change in Course Plans in the combined-sex analysis as well as those for each sex. This means that changes in Feelings completely track changes in Course Plans. We have seen earlier that linear changes in Feelings consistently predict linear changes in Course Plans: here we note that quadratic changes also run in tandem. There is every reason to believe that intervention programs aimed at influencing students' feelings about mathematics will also succeed in changing their preference for mathematics courses.

SES also predicts quadratic change in Course Plans, but only for boys. The positive coefficient implies that lower-SES boys have more of an inverted U-shaped curve than higher-SES boys. That is, the lower-SES boys increased substantially from sixth to seventh grade in their desire to take lots of mathematics, but increased less (or decreased) from seventh to eighth grade. This kind of jump from sixth to seventh grade could explain the connection of SES and level of Course Plans for boys in that the high mean of seventh-grade low-SES boys would have contributed to a higher overall level of Course Plans, so that this finding is useful in its explanatory significance. But it is hard to see why lower-SES boys should show such enthusiasm only in

TABLE 4.7
Separate Prediction Results for Males and Females Using Three Years of Data

	GRADE 6/GRADE 7/GRADE 8						GRADE 9/GRADE 10/GRADE 11					
	Level of Course Plans		Linear Change in Plans		Curved Change in Plans		Level of Course Plans		Linear Change in Plans		Curved Change in Plans	
	M	F	M	F	M	F	M	F	M	F	M	F
Level of Feelings	.26***	.25***					.18**					
Linear Feelings			.24***	.32***								
Curve of Feelings					.37***	.28***						
Level of Usefulness												
Linear Usefulness				.03**								
Curve of Usefulness												
Ability							.31**	.31**				
SES		−.10**						.11**				
R²	.43	.40	.22	.40	.36	.31	.45	.40	.13	.06	.07	.08
Number of Students in Analysis	143	128	144ᵃ	128	143	128	75	135	75	135	75	135

**p ≤ .01
***p ≤ .001
ᵃOne child for whom we lacked year two data on Course Plans was not included in the level or curve analyses, but was included in the linear change analysis.

seventh grade. Perhaps the content or the teaching of seventh-grade mathematics was especially enthralling to these boys, or English especially displeasing. It is hard to see how three school systems could manage such parallels, but it is certainly possible.

The following are our conclusions for the prediction of Course Plans for the younger students:

1. The Feelings measure is the best and most consistent predictor of the level of and changes in Course Plans for middle school students. Those who find mathematics easy and fun and enjoy their classes are more likely to want to take mathematics over humanities courses. Those whose enjoyment increases over time are likely to change similarly in Course Plans. Thus, educators planning intervention programs for middle school students should consider programs aimed at increasing students' pleasure in the study of mathematics. Changes in plans for involvement in mathematics can be expected to accompany changes in attitudes.

2. Usefulness does not enter as a predictor of level of Course Plans in any analysis, but it does predict linear change in Course Plans for girls. Thus, in considering intervention strategies, educators might plan to talk with girls about the ways mathematics can be useful to their everyday lives and work lives as adults.

3. Teacher Support and a perception of Mathematics as a Male Domain never enter as predictors of participation. At least in the way they are measured in this study, these variables do not seem relevant to the problem at hand.

4. Ability and SES do not have important roles in predicting Course Plans, though they do enter some of the prediction equations. Because they enter the prediction of level of Course Plans with negative coefficients, we can summarize their effects by saying that girls and boys in middle school have not concluded that their placement in low tracks for mathematics or their coming from low-SES homes might be detrimental to their participation in higher mathematics.

5. Even when we take account of individual differences in Feelings and all of the other variables discussed above, Gender predicts level of participation for seventh and eighth graders. By this age girls desire to take less mathematics than boys—and our analyses have not pinpointed any other explanation of this than Gender.

High School Students

Two-year analyses. In the analysis of level of Course Plans for the tenth and eleventh graders (see Table 4.4), a different set of predictors act as significant than was true for the younger children. The Feelings measure is irrelevant in predicting Course Plans, but Ability, SES, and Gender are very significant. Therefore, the following six conclusions seem appropriate:

Feelings do not have the same role for the older students as for the younger. The reason for this may be the differences in the measures of Course Plans. In any event, these attitudes toward the subject matter of mathematics do not predict the number of courses high school students say they will take in mathematics departments, although they do predict younger students' preferences for mathematics courses over humanities courses.

Usefulness and the three measures of the social milieu (Teacher Support, Math as a Male Domain, Distance Between Self and Mathematician) are unimportant in the prediction of Course Plans for high school students, as they were for middle school students. These indices of the social milieu, at least as they are measured here, do not seem to influence Course Plans.

Two significant predictors for the older group are Ability and SES. Everything else being equal, high-ability students plan to take more mathematics courses, as do students in higher socioeconomic classes. This is very different from the results for the younger students.

Gender also predicts level of Course Plans for high school students. The number in the table implies that boys and girls of equal ability and SES with equivalent attitudes toward mathematics do not enroll in the same number of high school mathematics courses. Boys take .34 more courses than girls. In a possible range of 0 to 3 courses, this is a considerable difference.

The only difference between the separate-sex analyses (see Table 4.5) and the combined-sex analysis is that the SES finding for the group as a whole is only repeated for girls. The general SES finding as stated above needs to be modified:

Girls from higher SES groups are more likely to want to take a large number of high school mathematics courses, but boys' choice of courses is not as strongly affected by SES. One reason for this difference may be that low-SES girls, having decided that a business or secretarial track is more desirable, stopped taking mathematics when it became optional. Low-SES boys may not have stopped mathematics for lack of clear options or because the available options (for example, carpentry, drafting) require an additional year or two.

Interestingly enough, none of the variables which we posed as possible predictors proved to be significant in our prediction of *change* in Course Plans for all students from the tenth to the eleventh grade. This means that we know how to figure out who among the high school students will want to take a lot of high school mathematics (highly able males and some high-SES highly able females), but we cannot predict who will make changes in Course Plans from one year to the next. Our separate-sex analyses were of no more assistance.

For those people interested in intervention programs for high school students, we can offer little advice. Students seem to have typed themselves as "in" or "out" of mathematics according to their Ability and SES. Changing their Feelings toward the subject, their knowledge of its Usefulness, or even their social milieu cannot be expected to change their Course Plans. At least, that is what these results would predict.

Three-year Analyses. Once again, since the major themes in these analyses were identical to those found in the two-year analyses, we will only report deviations from the latter results. Two of these occurred (see Tables 4.6 and 4.7). First, Gender was not a significant predictor of level of Course Plans in the three-year analysis for the high school students, although it was significant in the two-year test. Second, Feelings did not enter the two-year analyses in any significant way, but did enter for boys in the three-year analysis of level of Course Plans. Our changes in conclusions are the following:

The lack of significance of Gender in the three-year analysis is puzzling. The effect is quite large when only the tenth- and eleventh-grade scores are in the equation, but is not at all significant when the ninth-grade scores are added. With the middle school students we appealed to a lack of sex differences in Course Plans and most predictors for the sixth graders, but that is not the case with ninth graders, for whom many sex differences exist. We can offer the suggestion that the ninth graders (tested in the fall) had little idea of high school requirements for mathematics and were random in their answers, whereas the older students were knowledgeable and consistent.

The Feelings measure has a much weaker role for high school students than it does for middle school students. It has no role for girls, but does have a small predictive role for boys in the analysis of level of Course Plans. Boys who want to take more mathematics have more positive feelings about the subject, but even for this group changes in Feelings do not predict changes in Course Plans. We cannot offer a good prognosis for intervention programs that aim to change high school students' attitudes, as we have no evidence that such changes would affect Course Plans.

The analyses of the quadratic change scores indicated that none of our variables was successful at predicting significant quadratic changes in Course Plans. As we found with the linear change analyses of high school students, nothing safely predicts the changes they make in their number of planned courses in mathematics.

The following are our conclusions for the older students:

1. The Feelings measure is not a consistent predictor of Course Plans for high school students. In particular, there is no evidence that programs designed to change Feelings would, even if they succeeded in that aim, also change students' plans to take optional high school mathematics.

2. Usefulness and all three measures of the social milieu also seem unrelated to Course Plans. We have no reason to believe that attempts to influence these areas will result in changes in Course Plans.

3. Ability is a consistent predictor of level of Course Plans, implying that highly able students of both sexes intend to take more high school mathematics courses. It seems that high school students have typed themselves as able or unable to do mathematics, and we have no indication from the results of any changes in that opinion over time.

4. Socioeconomic status is also a principal predictor of Course Plans for girls. High school girls from higher-SES families intend to take more mathematics, probably so that they will be prepared to enter the colleges of their choice.

5. Gender entered to predict Course Plans only in the two-year analysis. That is, even after we have accounted for differ-

ences in attitudes, ability, and SES, boys in the tenth and eleventh grades plan to take more mathematics courses than girls, and no measures in our study can explain that difference. But the ninth-grade scores for girls and boys are not divided in this same fashion, and when they are included in the analyses, overall sex differences are mitigated. Ninth graders are still unsure about the number of high school mathematics courses they will take, so educators may be able to convince them that many would be useful.

SUMMARY

There is a tremendous difference between younger and older students in the sorts of variables that predict Course Plans and, therefore, in the tacks educators might choose to take in convincing them to want to take more mathematics and to enroll in more high school mathematics courses. For the younger students, Feelings predicts preferences for mathematics courses even to the point of tracking linear and quadratic changes in Course Plans. These students seem open to and ready for an intervention program, and we have an idea of how to design one. In addition, younger girls seem attentive to the Usefulness of mathematics as a reason to alter their Course Plans. This should serve as a second hint for the planning of intervention strategies.

By the time students are in high school, however, these glowing prospects have dimmed. The relevant predictors of Course Plans are Ability and, to a lesser extent, SES; both are variables which are not susceptible to much change from external sources. In trying to predict changes in participation over the three years of the study, we could find no variable of use. These students seem to have decided to take or avoid optional mathematics, and it is not immediately clear how to change their decisions.

NOTES

1. A contingency coefficient was used for these comparisons since both variables have only four levels and there were many ties.
2. We also asked high school students to record the actual courses they had taken by checking the appropriate names from a list of courses. The hope was that we could get a measure of the sophistication of their mathematical knowledge. The results were not usable. Many students checked course sequences, such as Essential Mathematics, Business Mathematics, Calculus, which we knew were incorrect. A thorough check of student records in one high school (the smallest) showed that nearly half of the students had checked incorrect sequences, even when such sequences were plausible. The problem was especially exacerbated by the fact that two school districts offered an Algebra I sequence that

lasted two years and was called Algebra IA and Algebra IIA. This was usually confused on the questionnaire with the traditional Algebra I and II. Another high school offered a joint Algebra/Geometry course and students didn't know what to check. The number of errors and the prohibitive amount of time required to check each question against student records made these data unusable.

3. It should be noted that this SES measure is limited in scope. Many researchers prefer to combine indices of parental occupation, education, and income to describe SES, and some use a series of questions about articles present in the home (e.g., daily newspaper, television, video-cassette recorder) to deduce SES. We asked students to report both parents' occupations and their educational attainments, but chose to use the occupation of the head of the household, because so many children did not know how far their parents had gone in school. We could check occupations through school records, but could not check educational status.

4. We used principal components factor analyses with varimax rotations for each of the factor analyses reported in this section.

5. A note is perhaps necessary concerning the reduction of the Support/No Support from Others scale to the items about teachers. First, many students reported to the testers that they thought encouragement from peers was equivalent to willingness to let homework be copied. Since this wasn't the meaning intended by the test constructors, these items were eliminated. Second, the parent items did not load systematically with each other or with any other items, so we chose not to use them as a scale. Further work is needed to determine how to solicit information about encouragement from students.

6. In the language of statistical analysis, we constructed orthogonal polynomials to accomplish this redefinition. These polynomials are used in analyses of variance when a specific order of means or other pattern of results is to be examined. The methods consist simply of constructing new variables by multiplying the old scores by predetermined weights. For the cases we have here, the weights are simply:

	Year 1	Year 2	Year 3
2-Year Weights			
Level		+1/2	+1/2
Change		−1/2	+1/2
3-Year Weights			
Level	+1/3	+1/3	+1/3
Change	−1/2	0	+1/2
Curve	+1/6	−1/3	+1/6

Where the base variables, the three years of Course Plans, have the same variances and intercorrelations, the new component variables will be uncorrelated and will give independent analyses.

7. We used ordinary least squares regressions, entering all variables simultaneously.

8. B-coefficients significant at the .05 level do not appear on these tables. In most cases where a coefficient was significant at this level but not at the .01 level, the coefficient was not significant for the comparison sample, and we felt that inclusion of the results would present an inaccurate picture.

9. This interpretation of the results from the regression equations is clearly causal in nature. While we have chosen to emphasize one direction of causality, we certainly acknowledge the possibility of the reverse situation. The reverse possibility is discussed explicitly in Chapter Five.

πϪπϪπϪ **5** πϪπϪπϪ

Suggestions for Changes

In the previous chapters we reviewed results from our three-year longitudinal study of sixth through twelfth graders. We will now summarize those results and consider their implications. The major findings can be summarized as follows:

1. For both sexes there is a decline in plans for involvement in mathematics from the sixth to the twelfth grade.

2. Concomitantly, there is a clear and widespread decline in positive attitudes toward mathematics. This can be translated into an increase in the perceived difficulty of the field and a decrease in both the perceived usefulness of mathematical knowledge and the degree to which the study of

mathematics is seen as creative, interesting, and fun. In high school there is also a decline in the level of support students feel they are given for their study of mathematics.

3. Such increasingly negative attitudes do *not* characterize all school subjects. Students' attitudes toward English, for example, remain constant or become more positive as students progress through school.

4. Boys' and girls' attitudes differ significantly with regard to the difficulty of mathematics, the degree to which it is anxiety-provoking, and the degree to which it is a male domain. Girls see mathematics as more difficult than do boys, as more anxiety-provoking, and as more clearly open to both sexes.

5. The feelings of sixth through eighth graders toward mathematics and the gender of seventh and eighth graders strongly predict the level of students' preferences for mathematics over humanities courses, and changes in feelings strongly predict changes in course preferences. For middle school girls, changes in the degree to which they see mathematics as useful for their future lives also predict changes in course preferences. For ninth through eleventh graders, level of participation in mathematics is associated with ability, socioeconomic status, and, for the older two grades only, gender. No factor, however, predicts change in participation.

The first three findings concern changes in students' ideas about mathematics and their plans for involvement in it. The seriousness of these changes and the apparently negative way in which many students view mathematics should be proof that there is a real problem here for educators. We are not training students sufficiently for the job market they will face upon leaving high school, nor are we adequately preparing many of them for the college courses they must take to earn an economically satisfying wage. In addition, we are not educating enough students in the art and science of logical thinking, and we are not providing many with pleasing classroom experiences that might prepare them to change their direction later on and take more mathematics. Too many students are saying that they dislike the experiences they have with mathematics so much that they just do not wish to have anything further to do with the subject. As a consequence, we are hindering them in the short term by not preparing them for the quantitative job market, and in the long term by not giving them the positive attitudes which might help them later in a transition to a more mathematically oriented job.

The fourth finding shows that the problem of negative attitudes and lack of desire to take many mathematics courses is especially acute for girls. They desire to take fewer mathematics courses than do boys as early as sixth grade. In any solution that we may propose for bringing students back into mathematics, we must pay attention to the special severity of the problem for girls.

The final summary finding is the most important for suggesting the directions our efforts may take to convince students that mathematics is interesting and necessary for their lives. The strong relationship of Feelings to Course Plans for the sixth to eighth graders suggests that one strategy is to attempt to alter Feelings so that Course Plans will change in turn. For middle school girls, the results imply that teachers or special programs could discuss careers in mathematics and uses for the subject matter in one's everyday life as an adult. For high school students, the results do not indicate obvious strategies for change because students seem already to have typed themselves as "good" or "bad" in mathematics, and we have no hint of how to change these opinions. We must therefore be creative in our strategies for working with these students to open the necessary doors.

We propose five possible strategies, some for sixth through twelfth graders, some for a subset of that grade range:

1. *Create a relaxed atmosphere in mathematics classrooms.* This is another way of saying: change students' feelings about mathematics, but aim at the core of these feelings, their reaction to the mathematics classroom. For middle school students this strategy may be expected, if successful, to change students' feelings and also to change their course plans. For high school students we cannot expect that it will change course plans, but it will certainly improve the quality of the day-to-day lives of the students.

2. *Present information to students on mathematical careers.* Our evidence suggests that if presenting such information succeeds in changing middle school girls' notions about the usefulness of mathematics, it will also change their course plans. While we have no similar evidence of this effect for sixth- to eighth-grade boys and high school students, we can argue that such an effort would still have benefits. For

example, few of the students tested had any knowledge of the uses of advanced mathematics outside the classroom. We were not comparing students with such knowledge to those without in our statistical tests, but perhaps such a comparison—which would be possible after an exploration of career options—would show that knowledge of concrete uses for mathematics does influence course plans for everyone.

3. *Devise special programs for parents, teachers, and guidance counselors on the usefulness of mathematics.* Our results showed that parents and teachers generally encourage students in their study of mathematics. While we cannot explain from our results how this encouragement works, we can at least suggest that if the significant people in a student's world all tell him or her that mathematics is a necessary subject to study, the pressure will be clearly placed on that student to keep trying.

4. *Introduce new nonremedial mathematics courses into the curriculum.* For many students in the middle school, making the mathematics class less formal should be sufficient to encourage them to continue their study of mathematics. But for some sixth to eighth graders and for high school students, it may be productive to develop new courses in addition to changing the atmosphere of regular classes. These new courses could be similar to the traditional courses, but with a different twist. For example, a teacher could try teaching algebra through the medium of social science experiments and statistics or the use of computers. Or perhaps it would be useful to introduce students to nontraditional areas of mathematics, such as non-Euclidean geometry or group theory, in interesting ways. The idea is that the success students would experience in these new courses might change their notions of their own ability, actually increase their mathematical knowledge, and thus motivate them to re-enter the regular mathematics curriculum or take more courses in mathematics than originally planned.

5. *Require four years of high school mathematics.* This study has pointed out the necessity of students learning mathematics well. It has also shown that there are relationships between some attitudes and course plans. Earlier strategies for increasing participation have been based on one interpretation of the results—that changes in attitudes will lead

to changes in course plans. This strategy is based on the opposite interpretation—that changes in course plans will change attitudes, and presumably, better prepare students for the world they will enter after graduation. Quite apart from the results, this strategy can be seen as solving the issue in one blow without the use of psychological research.

In the following sections we will describe in more detail each of these strategies for getting students more involved in mathematics, giving concrete examples of how each strategy might be implemented.

CHANGING THE MATHEMATICS CLASSROOM

The idea we would like to explore in this section is that if we can define the factors in the classroom environment that students dislike and change them, students and teachers alike will enjoy the experience more and students will be encouraged to take more courses. Let us first say this does not mean pandering to students. The intention is to teach the same amount of material in a mode that is comfortable for the teacher, but also pleasant for students.

In Chapters Three and Four we discussed the things students liked and disliked about the subject matter of mathematics and mathematics classrooms. We found that students really enjoyed classes which were interesting, those in which they understood the material, and those where they felt comfortable with their fellow students and the teacher. They also liked knowing that the material they were learning would be useful to them, although this consideration seemed less important than the atmosphere of the classroom.

The problem is that as students grow older, particularly in high school, they say they enjoy fewer classes. They feel less and less that they are in control of the material. They often cannot get good grades, and they don't see much use for the subject outside of the classroom. Not only do they find the subject difficult, but also distinctly anxiety-provoking. As each year goes by, more and more students do not feel free to ask questions in class; they see mathematics as a series of tests in which their egos are at stake, and they read the increasing formality of classroom sessions as an invitation to look stupid in front of friends and teacher—rather than as an invitation to learn. But not every subject is viewed this way by students.

Consider for a moment another result from this study. We asked students to rate the degree to which they liked or disliked a whole set of school subjects: biology, chemistry, English, history or social studies, languages, mathematics, and physics. The relative rankings of each of the subjects are presented in Table 5.1. In the sixth grade students placed mathematics high on their list of favorite courses, second only to science, which for children of this age probably meant biology. In the seventh grade, students ranked mathematics second only to chemistry, and in the eighth grade it followed languages and biology. By the ninth grade, however, mathematics lagged behind both English and history/social studies and it never recovered its position of prestige above the major humanities. The suggestion, then, is that we can borrow ideas for interesting classes from the subjects which students have reported liking best. We can take the lead from mathematics classes in the middle school, for instance, and from English classes in high school.

Borrowing from high school English classes seems an interesting tack to explore first. Particularly in high school, these classes are seen as more enjoyable than mathematics classes, more creative and fun, and as providing more useful skills. But why should this be? It is a very rare high school student who discovers new interpretations of Shakespeare in the course of writing a paper on a play, or writes about an emotion that no one else has experienced when he or she writes a short story. If the high school mathematics student is not inventing the binomial theorem, neither is the high school English student discovering new ideas and interpretations in literature. Why then should these experiences feel so different?

The answer seems to lie partly in teachers' attitudes and partly in students' feelings. Students report that English teachers seem to make them feel that they have done something wonderful when they put together a coherent analysis of a play or write a humorous story. They say that English teachers encourage them to talk about their ideas, and reward them with a grade that says they have thought well and come to meaningful conclusions. Mathematics teachers, on the other hand, are more likely only to reward a *right* answer—for what is a *meaningful* answer to a typical algebra problem? It is even common for teachers (and students) to be annoyed at the really bright student who finds alternative ways of proving a theorem or solving a problem. He or she disturbs the flow of the class, and the time spent in the diversion means that less is accomplished during the period. Rather than being appreciated for his or her originality, the student is often cut short when presenting an explanation.

TABLE 5.1

Relative Rankings of Liking of School Subjects

Grade 6	Grade 7	Grade 8	Grade 9	Grade 10	Grade 11	Grade 12
Science[a]	Chemistry	Languages	Biology	English	English	English
Mathematics	Mathematics	Biology	English	Biology	History	Biology
Languages	Languages	Mathematics	History	History	Biology	History
English	Biology	English	Mathematics	Mathematics	Mathematics	Mathematics
History	English	Chemistry	Languages	Languages	Languages	Languages
	History	History	Science[a]	Chemistry	Chemistry	Science[a]
	Physics	Physics		Physics	Physics	

[a]In the first year of the study the Science classification was used to encompass biology, chemistry, and physics for the sixth graders. Science was a combination of the ratings of chemistry and physics only for the ninth and twelfth graders.

In a mathematics class it is easy to emphasize detail, memorization, and the importance of figuring out the one right answer. Teachers, laboring under a long tradition, may never explicitly present underlying mathematical principles or the ramifications of a theorem. Whether this is because they don't see the importance of these facts in lending meaning to daily assignments, or because they assume that students will have arrived at such principles themselves by induction from daily assignments, the effect is often that many students are left in the dark. This, in turn, contributes to students' negative attitudes.

For their part, students approach mathematics classes with a very different set of attitudes and expectations from those they take to their English classes. Many students report that they have given up trying to understand mathematics—to integrate ideas or recognize general principles. Rather, they memorize a few rules —which they take to be random and arbitrary—before each test. English, on the other hand (beyond grammar, spelling, and punctuation), seems to be the stuff of human experience, amenable to thought, intuition, and feeling.

We should note in passing that we are in the midst of a back-to-basics movement in English as well as in mathematics. Parents want their children to know how to do computations, to write grammatically correct English, and to get high scores on standardized tests. As educators, we know there is more to learning mathematics and English than getting high scores on such tests, but we have a tough row to hoe to convince parents and school officials that our methods are sound if we cannot deliver on test scores as well as teach students to perform deeper, more intellectual processes, such as learning to think, discerning a false argument, or inventing a new approach.

The comparison between high school English and mathematics classes provides us with suggestions for directions to take in making classes more enjoyable:

Encourage students to express their own ideas.

Ask questions which do not have only one right answer.

Reward interesting approaches to solving problems even if they do not lead to the right answer.

Talk about the reasons students are learning about a topic in terms of how this topic fits into mathematics and how it helps one to understand real-world phenomena.

Translate theorems into common parlance whenever possible.

Summarize topics frequently so that students know how today's lesson is related to yesterday's and tomorrow's.

Repeatedly demonstrate that mathematical algorithms are not random and arbitrary, but are understandable and logical.

Without changing the substance of their courses, teachers can change elements of their style and of the atmosphere they set in the classroom, and trust that such changes will have a significant impact on students.

From a look at a successful sixth- and seventh-grade mathematics class it is possible to come up with other suggestions for introducing quantitative ideas. In this class the teacher decided to experiment with her lesson plan in the geometry section, and introduced a problem about building a house. In order to teach her students about the concepts and skills involved in calculating area, she presented them with the task of designing a five-room house to be built on a lot of specified acreage, while leaving a certain minimum area for a yard. She allowed her students to explore problems with perimeters by presenting the student architects with a maximum amount of lumber which could be bought to build the structure's walls. She asked that students work in small groups or pairs, and she moved about the room and visited with different groups to answer questions and offer suggestions. Several class sessions were used for the project, giving students time to formulate their ideas, discuss the project among themselves, and work through the intricacies of problems as they arose. Additional problems, or miniprojects, such as designing a "modern" house with nonrectangular rooms with certain area and perimeter restrictions, were available for students who finished the assignment more quickly than others. Toward the end of the project, the teacher asked the groups to present their floor plans to the class to see the diversity in designs and to discuss the different conceptions of the house. For special interest she invited an architect and a contractor to critique the designs and talk about the importance of geometry in their work. Then she summarized the principles students had used intuitively by translating them into mathematical terms.

Why might such a lesson be seen favorably by students? First of all, most students were intrigued by the problem and thought the construction of the house plans was fun. Based in a concrete problem, the material seemed to be more easily understood than abstract lectures on geometry, and the approach made problems in geometry seem more practical and relevant to everyday life. The cooperative work in small groups made the atmosphere in the classroom more informal and less competitive, and allowed each student to receive more personal help and attention from his or her teacher. Since the project grade was awarded jointly by class members, the teacher, the architect, and the contractor, students felt that the evaluation system was not as anxiety-provoking as an examination or quiz would have been at the end of a unit. Students weren't singled out, but were assessed as a group. Each student could contribute to the decision about his or her own grade, so students maintained some control over their evaluations. The diversity in floor plans among the working groups illustrated that there could be more than one correct process and solution to a mathematical problem, and that geometry was a subject area in which students could express their own creative ideas. The follow-up talks by the architect and contractor made geometry seem relevant to students' everyday lives, as well as to possible careers.

Needless to say, these kinds of lessons require a lot more time on the part of the instructor and probably the students. It is not easy to dream up interesting problems which will capture students' interest. And the extra time spent planning and reviewing these sorts of assignments means more work for teachers; they cannot follow the text, assign the usual series of homework problems, and discuss those problems the next day in class. They have to think about presenting the ideas in a creative way that students will recognize as useful, and that ties the material to concrete knowledge which students already possess. They must also devise a method of presentation which invites questions and student involvement and will, in turn, assure that students do indeed understand the material.

It may be that such radical deviations from the typical high school classroom are not possible, and that such projects are appropriate for sixth-, seventh-, and eighth-grade mathematics classes, but not for high school. Whether that be the case would depend on the specific teacher and the requirements of the particular school system. In the event that such activities are not possible, it should still be practicable to make the changes suggested earlier based on the success of high school English teachers. Anything that reduces the tension frequently present in mathematics classes would benefit students and, indirectly, their teachers.

DESIGNING SPECIAL PROGRAMS FOR STUDENTS
ON CAREERS IN MATHEMATICS

The notion behind introducing a special program on careers is that the information gained by students about mathematical careers will motivate them to pursue mathematics courses, even if they do not particularly enjoy the experiences. If they can see why they should study mathematics—because it will be useful to them in the future—then they will keep working on a hard problem, and keep facing the difficulties they experience with mathematics in their regular class assignments.

One particularly strong rationale for introducing this kind of a program is that data have shown that girls in the sixth to eighth grades will be responsive to such information and will change their course plans if convinced that they need to take more mathematics. And we know that this enrollment problem is much more serious among girls than among boys.

A school system or a teacher could inaugurate a program by asking students to read any of a number of publications now on the market, for example, *I'm Madly in Love with Electricity* (Kreinberg, 1977) and *Math Equals* (Perl, 1978);[1] by collecting and distributing information from the guidance office on requirements for careers and openings for women; by doing research in a local library about famous women in mathematics and the sciences and discussing this with students; or by appealing to several outside organizations for help. The Math/Science Network in Berkeley, California, for example, sponsors career conferences for students to hear in person about opportunities in the quantitative job market. Project WAM (Women and Mathematics) of the American Mathematical Society brings women in the sciences into the classroom to talk with students about their experiences. Project WITS (Women in Technology and Science) of the Massachusetts Institute of Technology and Project EQUALS of the Lawrence Hall of Science (University of California at Berkeley) offer teachers more information on career opportunities in quantitative areas and how to encourage girls to enter them.[2]

If a school system could manage a systemwide program, it would be possible to organize a career weekend, for example, and women in different careers could be asked to talk with students. Or a series of career talks could be arranged for special periods in the day or after school. It would be sensible to have school guidance counselors and mathematics teachers pull together such a program, if possible, with the help of the organizations which now exist and the information they can provide about careers.

DESIGNING PROGRAMS FOR PARENTS, TEACHERS, AND GUIDANCE COUNSELORS

It may be that part of the reason students are not aware of the many uses of mathematics is that teachers, parents, and guidance counselors do not have a wealth of information about such uses. A good way to ensure that students hear about their need for mathematics is to make sure that the important figures in their environment are aware of the need and are armed with enough information to persuade the students that it exists.

The kind of programs mentioned above aimed at informing students about career options in mathematical areas, might be appropriate as special programs for parents, teachers, and guidance counselors, as well as for students. If such programs were to take place in the evening or on a weekend, adults could attend with their children. Or it might be advantageous to run seminars exclusively for adults. This could be done by having speakers come to talk with teachers and guidance counselors on a special day—the system's annual Teacher Day, for example. In addition, the local Parent and Teachers Association or similar organization might choose to sponsor a session or two on career options. Such sessions could inform parents about options for their children *and* for themselves.

INTRODUCING NEW MATHEMATICS COURSES

If one reviews the vast amount of psychological literature on therapy or behavior change, one common theme emerges: A method which convinces someone to behave in a desired manner and then rewards that person for the new behavior is more successful at retaining the continued behavior than simply talking with the person about his or her problems. An example may help to clarify this generalization.

Many children are afraid of dogs, and need to learn to approach dogs properly. Bandura (1968) developed a method for coping with children who were afraid of dogs which started with having the child come to a party with a lot of other children where there was a small dog in a play pen in one corner of the room. The dog could not get out, so was not an immediate threat to the frightened child. In the course of having fun at the party, the child would be encouraged by his or her friends to go look at the dog. One friend might pat the dog, and an adult might point out that the tail was wagging and that was a good sign. The child would be encouraged to pat the dog on the head, but not forced to do so. The children around the pen would start a conversation about their dogs and how they treated them and played with them so that dogs would come to be acceptable partners at a party. As the "treatment" progressed, adults would continue to provide the child with successful experiences with dogs. In no case would the child be forced to deal with a dog in an overly frightening situation; rather, adults would gradually build the series of experiences so that the child eventually interacted responsibly with dogs. The desired result is for the child to experience success and to replace his or her fear of dogs with the knowledge of how to deal with animals.

The point of this example is that the same sort of treatment can work to reduce negative feelings about mathematics. If teachers can give students a relaxed, pleasant classroom and help them have positive experiences with the subject, their negative feelings may be ameliorated, and they may even be convinced to take more mathematics. However, for those students who are far down the path of dislike for mathematics, a somewhat different tack may be needed to convince them to approach the subject again. It might be necessary to start outside of the regular curriculum by offering new courses dealing with nontraditional material.

These new courses could be of two types: they might introduce the same ideas as are introduced in the regular curriculum, but through the use of a different medium; or mathematical ideas which are not generally discussed in middle and high school could be covered. In the first instance it might be possible to introduce a computer course into the curriculum to teach algebra by instructing students to program a computer. Perhaps one of the social science courses in high school could be designed to review arithmetic and elementary algebra by conducting psychological experiments and analyzing the data. An exciting consumer mathematics course could be concerned with shortcuts in working simple calculations, and algebraic manipulations could be introduced along the way. The rationale for such courses is that students did not gain a sufficient understanding of arithmetic and algebra the first time around, but their attitudes toward mathematics are so negative that they won't gain a much better understanding a second time around either. From the first day forward, the new course must be different enough and students must be successful enough in it to overcome their negative attitudes.

109

A sure way to make this effort fail would be to label a new course remedial. Implying that anyone is less able than other students, or inferior, or different would have the same effect as telling the child who is afraid of dogs that he or she is sick mentally. It would discourage further participation in mathematics and would probably ensure that students paid only minimal attention to the effort. The courses must be oriented toward encouraging students in mathematics and providing them with successful experiences which will engender further positive experiences.

The second revision of the curriculum could be the introduction of topics that students may not immediately recognize as mathematical, the intent being to convince students that mathematics can be innovative and exciting. Such courses should encourage students to play with mathematical ideas, and should not have the usual rigid set of topics and exercises to be accomplished within a certain period of time. The goal of the courses would be to get students to think in creative ways about mathematical problems and to convince them that their own ideas are needed in mathematics.

One example of such a successful course has been taught at Wellesley College for the last several years. Its topics include the place of symmetries in art and architecture or in the ringing of changes on bells (to introduce group theory), the planning of postal routes (to talk about graph theory), and the question of population growth (to discuss the basis for calculus), among others. Students work together in class to solve very concrete problems. They propose solutions and test them out for themselves. They develop the theory in each of these fields, and the instructor makes formal ties between their discoveries and traditional theorems. Students come to realize that they are working with tough mathematical problems with which experts in the field are still struggling. Many students are initially skeptical about whether they are even taking mathematics, but come out of the course thinking the discipline is really very creative, applicable outside of the traditional areas, and that they would like to pursue the subject.[3]

There are probably an infinite number of ways to introduce students or reintroduce students to mathematics as a pleasure to study, as many as there are teachers interested in doing so. By setting up new courses in the curriculum or altering course objectives to include changing students' attitudes, students can be pulled back into mathematics without forcing them into the regular curriculum until they are convinced they should be there.

The drawback to instituting these ideas in many school systems is partly a financial one. With the increased costs of energy and decreasing student populations, such endeavors are hard to justify. The pressure from parents who want their children to know the "basics" does not include a pressure to encourage their children to *enjoy* the basics. So if teachers, principals, and guidance staff are convinced that the curricula discussed above are too valuable to be dismissed, they must develop a strategy for passing the word along to parents and the school board. Outside funding from a business or a governmental agency would help; so too would a saturation campaign about the need for students to study mathematics to prepare themselves for the job market of the future, and the need for new courses to foster students' learning of the necessary skills.

REQUIRING FOUR YEARS OF HIGH SCHOOL MATHEMATICS

This last strategy, requiring students to take four years of mathematics, sounds simple since it implies that all students would graduate from high school better prepared in mathematics than many are now. Unfortunately, such a strategy raises a host of problems which make its adoption by most school systems unlikely.

First, without changing the regular curriculum by relaxing the classroom atmosphere or introducing new courses, a school system adopting this four-year requirement would be forcing mathematics teachers to teach students who hate the subject for an even longer period of time. It isn't very likely that these students would suddenly change their minds about mathematics during the extra years of required study. It is more likely that the additional years would make students' and teachers' lives more miserable.

Thus, this strategy would only be successful if accompanied by one or more of the other strategies suggested above. If all of them were combined into one system, it would place an interesting burden on a mathematics department to devise courses to meet the needs of students at all levels of ability and knowledge, and would envelop the staff in a difficult, though potentially creative, effort. In some sense this is the problem that English teachers have faced over the last few years. English is still required through all four years in most high schools, and English departments have responded to students' interests and abilities by changing their traditional curricula in ways responsive to student needs. For example, many school systems offer elective courses in the eleventh and twelfth grades, giving students the choice to take such pleasing courses as Mystery and Science Fiction, The American Novel, Writing Short Stories, and Script Writing for Television. Adding these enticements to the list of courses which includes Secretarial English and Composition I means that students can choose to review grammar, spelling, and punctuation or to explore more esoteric realms of literature and creative writing. Requiring four years of high school mathematics may force mathematics teachers to think hard about making their subject just as interesting. A better way of putting this is that such a requirement would give free rein to mathematics teachers to be creative within their discipline instead of being tied to the traditionally rigid and less rewarding curriculum.

One can argue that a new four-year curriculum with electives would be a boon for teachers and students alike. For once, teachers could be inventive in developing their mathematics courses. Students would have more control over the choices they were making, could respond to the creative and innovative curricula and the enthusiasm of their teachers in a positive manner, could enjoy mathematics more, and could learn what they need to know to enhance their futures.

Of course, as stated earlier, all of this requires money. The school system would have to hire more mathematics faculty to cover all of the old and new courses, and would have to buy new texts and materials. It would take a strong case and hours of argument to convince a system this change is wise, but a four-year curriculum might prove to be the perfect solution to the problem of preparing students adequately for the job market of the future.

CONCLUSION

At the beginning of this book we talked a great deal about the need for advanced mathematics or at least a thorough understanding of mathematics and a willingness to learn more as preparation for tomorrow's job market. In this chapter we have used the results of this study to suggest five potential solutions to getting more students into advanced mathematics courses. Each solution has its problems and demands a great deal of effort on the part of teachers and possibly other school personnel. But such efforts need to be made and can be made successfully if teachers work together to convince others of the need for additional mathematical training and the likelihood of success for their strategies.

We also opened this book by illustrating the serious need for the increased participation of girls in mathematics courses, since girls are dropping out of the field earlier than boys and will therefore have more trouble re-entering the field if they wish to do so, or getting better-paying jobs later on. The results of this study suggested that the same sorts of strategies for changing boys' minds about mathematics would work with girls, and that girls are especially inclined to respond to arguments about the uses mathematics could have for them in the future. School systems should pay attention to the especially serious need among girls for information about opportunities in mathematics and the sciences, and initiate programs which are convincing for their students. Girls need to know that it is very likely they will work for a good deal of their lives, even if they have children and stay married. They also need to know that high-paying jobs *are* available to them, and will be relatively easy to return to if they wish to take time off to have children—but only if they stay with mathematics in high school and prepare themselves for the opportunities.

As we all know from experience, telling someone what is good for them and having them act on the suggestions are two very different phenomena. Teenagers have much more on their minds than preparing themselves for some theoretical job market many years off, and the social activities of the present loom so large in their minds that they might not even hear the information we are supplying. Our solution must be to saturate their environment with the need for the study of mathematics. This means convincing parents and school staff alike of the need, and talking frequently to students about careers throughout middle and high school in every regular mathematics class and in special programs outside the regular curriculum. Encouraging students in mathematics—especially girls—should be a major undertaking for all educators in the 1980s, and it must begin today.

NOTES

1. Additional publications of interest include the following:
   "Bibliography on Careers in Mathematics and Related Fields," Society for Industrial and Applied Mathematics, 33 South 17th St., Philadelphia, PA 19103. 1976. Free.
   "Careers for Women in Mathematics," Association of Women in Mathematics, Department of Mathematics, Wellesley College, Wellesley, MA 02181. 1980. Free.
   "Careers in Mathematics," Mathematical Association of America, 1225 Connecticut Ave., N.W., Washington, D.C. 20036. Free.
   "Careers in Statistics," American Statistical Association, 806 15th St., N.W., Washington, D.C. 20005. 1974. Free.

"Career Mathematics: Industry and the Trades," Houghton Mifflin Co., 1 Beacon St., Boston, MA 02107. 1977. $6.99.

"Math: Who Needs It?" Counselor Films, Inc., Career Futures, Inc., 2100 Locust St., Philadelphia, PA 19103. $25 for film and cassette.

"Mathematics and My Career," National Council of Teachers of Mathematics, 1906 Association Drive, Reston, VA 22091. 1971. $1.75.

"Professional Opportunities in Mathematics," Mathematical Association of America, 1225 Connecticut Ave., N.W., Washington, D.C. 20036. 1974. $0.50.

"Professional Training in Mathematics," American Mathematics Society, P.O. Box 6248, Providence, R.I. 02940. 1976. $0.80.

"So You're Good at Math," Society of Actuaries, 208 South LaSalle St., Chicago, IL. 60604. Free.

"The Math You'll Need for College," Mathematical Association of America, 1225 Connecticut Ave., N.W., Washington, D.C. 20036. Free.

2. For information on the Math/Science Network write to:

> Ms. Nancy Kreinberg
> Director of Secondary Programs
> Lawrence Hall of Science
> University of California
> Berkeley, CA 94720

or

> Dr. Lenore Blum
> Director of Collegiate Programs
> Mathematics and Computer Science
> Mills College
> Oakland, CA 94613

or

> Ms. Joanne Koltnow, Coordinator
> Math/Science Resource Center
> Mills College
> Oakland, CA 94613

For information on Project EQUALS and other Lawrence Hall of Science Projects, contact Nancy Kreinberg as well. For details on WITS and WAM, the following addresses should be useful:

> Ms. Edith Ruina
> Women in Technology and Science (WITS)
> Room 20C–228
> Massachusetts Institute of Technology
> Cambridge, MA 02139

> Dr. Eileen Poiani
> Women and Mathematics Program (WAM)
> Department of Mathematics
> Saint Peter's College
> Kennedy Boulevard
> Jersey City, NJ 07306

3. For further information and a copy of curriculum units contact:

> Dr. Alice Schafer
> Department of Mathematics
> Wellesley College
> Wellesley, MA 02181.

Bibliography

Abrego, M. B. 1966. Children's attitudes toward arithmetic. *The Arithmetic Teacher* 13:206-8.

Adams, S., and Von Brock, R. C. 1967. The development of the A-V scale of attitudes toward mathematics. *Journal of Educational Measurement* 4:247-48.

Admissions Testing Program of the College Entrance Examination Board. 1977. *National Report on College-Bound Seniors.* Princeton, N.J.

Ahlgren, A., and Walberg, H. J. 1973. Changing attitudes toward science among adolescents. *Nature* 245:187-90.

Aiken, L. R. 1963. Personality correlates of attitude toward mathematics. *Journal of Educational Research* 56:476-80.

_____.1970a. Attitudes toward mathematics. *Review of Educational Research* 40:551–96.

_____. 1970b. Nonintellective variables and mathematics achievement: Directions for research. *Journal of School Psychology* 8:28–36.

_____. 1971. Intellective variables and mathematics achievement: Directions for research. *Journal of School Psychology* 9:201–12.

_____. 1972a. Biodata correlates of attitudes toward mathematics in three age and two sex groups. *School Science and Mathematics* 72:386–95.

_____. 1972b. Research on attitudes toward mathematics. *The Arithmetic Teacher* 19:229–34.

_____. 1972c. Language factors in learning mathematics. *Review of Educational Research* 42:359–85.

_____. 1973. Ability and creativity in mathematics. *Review of Educational Research* 43:405–32.

_____. 1974. Two scales of attitude toward mathematics. *Journal for Research in Mathematics Education* 5:67–71.

_____. 1976. Update on attitudes and other affective variables in learning mathematics. *Review of Educational Research* 46:293–311.

Aiken, L. R., and Dreger, R. M. 1961. The effect of attitudes on performance in mathematics. *Journal of Educational Psychology* 52:19–24.

Alexander, V. E. 1962. Sex differences in seventh grade problem solving. *School Science and Mathematics* 62:47–50.

Almquist, E., and Angrist, S. S. 1970. Career salience and atypicality of occupational choice among college women. *Journal of Marriage and the Family* 32:242–49.

_____. 1971. Role model influences on college women's career aspirations. In *The Professional Woman,* ed. A. Theodore. Cambridge, Mass: Schenkman.

Alpert, R., and Haber, R. N. 1960. Anxiety in academic achievement situations. *Journal of Abnormal and Social Psychology* 61:207–15.

American Mathematical Society. 1975. Nineteenth Annual AMS Survey, *Notices of the American Mathematical Society* 22:303–8.

Anastasi, A. 1958. *Differential Psychology: Individual and Group Differences in Behavior.* New York: The MacMillan Company.

Anastasi, A., and Schafer, C. E. 1969. Biographical correlates of artistic and literary creativity in adolescent girls. *Journal of Applied Psychology* 53:267–78.

Anderson, K. E. 1963. A comparative study of student self-ratings on the influence of inspirational teachers in science and mathematics in the development of intellectual curiosity, persistence, and a questioning attitude. *Science Education* 47:429–37.

Angrist, S. S. 1971. Personality maladjustment and career aspirations of college women. *Sociological Symposium.*

Anttonen, R. G. 1969. A longitudinal study in mathematics attitude. *Journal of Educational Research* 62:467–71.

Armstrong, J. A. 1979. A national assessment of participation and performance of women in mathematics. Paper presented at the Society for
Research in Child Development meetings, March 1979, in San Francisco, California.

Astin, H. A. 1967. Factors associated with the participation of women doctorates in the labor force. *Personnel and Guidance Journal* 46:240-5.

_____. 1968a. Career development of girls during the high school years.
Journal of Counseling Psychology 15:536-40.

_____. 1968b. Stability and change in the career plans of ninth grade
girls. *Personnel and Guidance Journal* 46:961-66.

_____. 1974. Sex differences in mathematical and scientific precocity. In
Mathematical Talent: Discovery, Description, and Development, eds.
J. Stanley; D. Keating; and L. Fox, pp. 70-86. Baltimore: The Johns
Hopkins University Press.

Astin, H. A., and Myint, T. 1971. Career development and stability of
young women during the post high school years. *Journal of Counseling Psychology* 18:369-93.

Austin, M. D. 1971. Dream recall and the bias of intellectual ability. *Nature* 231:59.

Bachman, A. M. 1970. The relationship between a seventh-grade pupil's
academic self-concept and achievement in mathematics. *Journal for
Research in Mathematics Education* 1:173-79.

Bachtold, L. M., and Werner, E. E. 1972. Personality characteristics of
women scientists. *Psychological Reports* 31:391-96.

Backman, M. E. 1972. Patterns of mental abilities: Ethnic, socioeconomic,
and sex differences. *American Educational Research Journal* 9:1-12.

Bandura, A. 1968. Modelling approaches to the modification of phobic
disorders. In *The Role of Learning in Psychotherapy,* ed. Ruth Porter,
pp. 201-17. London: J. & A. Churchill.

Barnett, R. C., and Baruch, G. 1974. Occupational and educational aspirations and expectations: A review of empirical literature. Unpublished
paper.

Bassham, H.; Murphy, M.; and Murphy, K. 1964. Attitude and achievement in arithmetic. *Arithmetic Teacher* 11:66-72.

Beardslee, D. C., and O'Dowd, D. D. 1960. College student images of a selected group of professions and occupations. H.E.W. Cooperative Research Project No. 562(8142).

_____. 1961. The college student image of the scientist. *Science*
133:997-1001.

Bem, S. L. 1974. The measurement of psychological androgyny. *Journal of
Consulting and Clinical Psychology* 42:155-62.

Bennett, S. N. 1973. Divergent thinking abilities—A validation study.
British Journal of Educational Psychology 43:1-7.

Berry, J. W. 1966. Temne and Eskimo perceptual skills. *International Journal of Psychology* 1:207-29.

Bickel, P. J.; Hammel, E. A.; and O'Connell, J. W. 1975. Sex bias in graduate admissions: Data from Berkeley. *Science* 187:398–403.

Biggs, J. B. 1959. Attitudes to arithmetic—Number anxiety. *Educational Research* 1:6–21.

Bock, D. R., and Kolakowski, D. 1973. Further evidence of sex-linked major-gene influence on human spatial visualizing ability. *American Journal of Human Genetics* 25:1–14.

Boswell, S. L. 1979. Sex roles, attitudes, and achievement in mathematics: A study of elementary school children and Ph.D.s. Paper presented at the Society for Research in Child Development meetings, March 1979, in San Francisco, California.

Brush, L. R. 1978. A validation study of the mathematics anxiety rating scale. *Educational and Psychological Measurement* 38:485–90.

————. 1979. Avoidance of science and stereotypes of scientists. *Journal of Research in Science Teaching* 16:237–41.

Callahan, W. J. 1971. Adolescent attitudes toward mathematics. *Mathematics Teacher* 64:751–55.

Campbell, N. J., and Schoen, H. L. 1977. Relationships between selected teacher behaviors of prealgebra teachers and selected characteristics of their students. *Journal for Research in Mathematics Education* 8:369–75.

Carey, G. L. 1958. Sex differences in problem-solving performance as a function of attitude differences. *Journal of Abnormal and Social Psychology* 56:256–60.

Carnegie Commission on Higher Education. 1973. *Opportunities for Women in Higher Education.*

Casserly, P. L. 1975. An assessment of factors affecting female participation in advanced placement programs in mathematics, chemistry, and physics. Washington, D.C.: Report to the National Science Foundation.

Castenada, A.; McCandless, B.; and Palermo, D. 1956. The children's form of the Manifest Anxiety Scale. *Child Development,* 27:317–26.

Cattell, R. 1958. The nature of anxiety: A review of thirteen multivariate analyses comprising 814 variables. *Psychological Reports* 4:351–88.

Centra, J. A. 1974. *Women, Men, and the Doctorate.* Princeton, N.J.: Educational Testing Service.

Channon, C. E. 1974. The effect of regime on divergent thinking scores. *British Journal of Educational Psychology* 44:89–91.

Chansky, N. M. 1966. Anxiety, intelligence, and achievement in algebra. *The Journal of Educational Research* 60:90–91.

Child, D., and Smithers, A. 1973. An attempted validation of the Joyce-Hudson scale of convergence and divergence. *British Journal of Educational Psychology* 43:57–62.

Cleveland, G. A., and Bosworth, D. L. 1967. A study of certain psychological and sociological characteristics as related to arithmetic achievement. *Arithmetic Teacher* 14:383–87.

College Placement Council. 1978. *CPC Salary Survey, March 1978*. Available through College Placement Council, P.O. Box 2263, Bethlehem, PA 18001.

Connor, J. M.; Serbin, L. A.; and Nosofsky, R. 1979. Visual-spatial skills and mathematics achievement. Paper presented at the American Psychological Association meetings, September 1979, in New York.

Cropley, A. J. 1966. Creativity and intelligence. *British Journal of Educational Psychology* 36:259-66.

_____. 1967. Divergent thinking and science specialists. *Nature* 215:671-72.

Cropley, A. J., and Field, T. W. 1968. Intellectual style and high school science. *Nature* 217:1211-12.

_____. 1969. Achievement in science and intellectual style. *Journal of Applied Psychology* 53:132-35.

Cropley, A. J., and Maslany, G. W. Reliability and factorial validity of the Wallach-Kogan creativity tests. *British Journal of Psychology* 60:395-98.

D'Augustine, C. H. 1966. Factors relating to achievement with selected topics in geometry and topology. *Arithmetic Teacher* 13:192-97.

Degnan, J. A. 1967. General anxiety and attitudes toward mathematics in achievers and underachievers in mathematics. *Graduate Research in Education and Related Disciplines* 3:49-62.

Dellas, M., and Gaier, E. L. Identification of creativity: The individual. *Psychological Bulletin* 73:55-73.

Dreger, R. M., and Aiken, L. R. 1957. The identification of number anxiety in a college population. *Journal of Educational Psychology* 48:344-51.

Droege, R. C. 1967. Sex differences in aptitude maturation during high school. *Journal of Counseling Psychology* 14:407-11.

Duckworth, D.; and Entwistle, N. J. 1974. Attitudes to school subjects: A repertory grid technique. *British Journal of Educational Psychology* 44:76-83.

Dutton, W. 1951. Attitudes of prospective teachers toward arithmetic. *Elementary School Journal* 52:84-90.

_____. 1954. Measuring attitudes toward arithmetic. *Elementary School Journal* 55:24-31.

_____. 1956. Attitudes of junior high school pupils toward arithmetic. *School Review* 64:18-22.

_____. 1962. Attitude change of prospective elementary school teachers toward arithmetic. *Arithmetic Teacher* 9:418-24.

_____. 1965. Prospective elementary school teachers' understanding of arithmetical concepts. *The Journal of Educational Research* 58:362-65.

_____. 1968. Another look at attitudes of junior high school pupils toward arithmetic. *Elementary School Journal* 68:265-68.

Dutton, W., and Blum. M. P. 1968. The measurement of attitudes toward arithmetic with a Likert-Type test. *Elementary School Journal* 68:259-64.

Easterday, K., and Easterday, H. 1968. Ninth-grade algebra, programmed instruction, and sex differences: An experiment. *Mathematics Teacher* 61:302-7.

Ekstrom, R. B.; French, J. W.; Harman, H. H.; and Dermen, D. 1976. *Manual for Kit of Factor-Referenced Cognitive Tests.* Princeton, N.J.: Educational Testing Service.

Elton, C. F. 1967. Traditional sex attitudes and discrepant ability measures in college women. *Journal of Counseling Psychology* 14:538-43.

Elton, C. F., and Rose, H. A. 1967. Significance of personality in the vocational choice of college women. *Journal of Counseling Psychology* 14:293-98.

Endler, N. S.; Hunt, J. M.; and Rosenstein, A. J. 1962. An S-R inventory of anxiousness. *Psychological Monographs* 76: Whole number 536.

Entwisle, D. R., and Greenberger, E. 1972. Adolescents' views of women's work role. *American Journal of Orthopsychiatry* 42:648-56.

Ernest, J. 1976. Mathematics and sex. *American Mathematical Monthly* 83:595-614.

Fedon, J. P. 1958. The role of attitude in learning arithmetic. *Arithmetic Teacher* 5:304-10.

Feldhusen, J. F.; Denny, T.; and Condon, C. F. 1965. Anxiety, divergent thinking, and achievement. *Journal of Educational Psychology* 56:40-45.

Fellows, M. M. 1973. A mathematics attitudinal device. *Arithmetic Teacher* 20:222-23.

Fennema, E. 1974a. Mathematics learning and the sexes: A review. *Journal for Research in Mathematics Education* 5:126-39.

_____. 1974b. Sex differences in mathematics-learning: Why??? *Elementary School Journal* 75:183-90.

_____. 1975. Spatial ability, mathematics, and the sexes. In *Mathematics Learning: What Research Says About Sex Differences,* ed. E. Fennema. Columbus, Ohio: ERIC Center for Science, Mathematics, and Environmental Education, College of Education, Ohio State University.

_____. 1976. Influences of selected cognitive, affective, and educational variables on sex-related differences in mathematics learning and studying. Paper prepared for National Institute of Education.

Fennema, E., and Sherman, J. A. 1976. Fennema-Sherman mathematics attitude scales: Instruments designed to measure attitudes toward the learning of mathematics by females and males. American Psychological Association Document MS1225. Available through Journal Supplement Abstract Service, American Psychological Association, 1200 Seventeenth St., N.W., Washington, D.C. 20036.

_____. 1977. Sex-related differences in mathematics achievement, spatial visualization, and affective factors. *American Educational Research Journal* 14:51-71.

Ferguson, L. R., and Maccoby, E. E. 1966. Interpersonal correlates of differential abilities. *Child Development* 37:549-71.

Field, T. W., and Poole, M. E. 1970. Intellectual style and achievement of arts and science undergraduates. *British Journal of Educational Psychology* 40:338–41.

Flanagan, J. C. 1976. Changes in school levels of achievement: Project Talent ten and fifteen year retests. *Educational Researcher* 5:9–12.

Flanagan, J. C.; Davis, F. B.; Dailey, J. T.; Shaycoft, M. F.; Orr, D. B.; Goldberg, I.; and Neyman, C. A. 1964. *The American High-School Student.* Pittsburgh; University of Pittsburgh.

Fox, L. H. 1975a. Career interests and mathematical acceleration for girls. Paper presented at annual meeting of the American Psychological Association, August 1975, at Chicago, Ill.

_____. 1975b. Mathematically precocious: Male or Female? In E. Fennema (Ed.), *Mathematics Learning: What Research Says about Sex Differences.* Columbus, Ohio: ERIC Center for Science, Mathematics, and Environmental Education, College of Education, Ohio State University.

_____. 1976a. The effects of sex role socialization on mathematics participation and achievement. Paper prepared for National Institute of Education.

_____. 1976b. Women and the career relevance of mathematics and science. *School Science and Mathematics* 76:347–53.

Fox, L. H., and Denham, S. A. 1974. Values and career interests of mathematically and scientifically precocious youth. In *Mathematical Talent: Discovery, Description, and Development,* eds. J. Stanley; D. Keating; and L. Fox, pp. 140–75. Baltimore: The Johns Hopkins University Press.

Fox, L. H.; Fennema, E.; and Sherman, J. 1977. Women and Mathematics: Research Perspectives for Change. Washington, D.C.: The National Institute of Education.

Garai, J. E., and Scheinfeld, A. 1968. Sex differences in mental and behavioral traits. *Genetic Psychology Monographs* 77:169–299.

Garron, D. C. 1970. Sex-linked recessive inheritance of spatial and numerical abilities, and Turner's syndrome. *Psychological Review* 77:147–52.

Garwood, D. S. 1964. Personality factors related to creativity in young scientists. *Journal of Abnormal and Social Psychology* 68:413–19.

Gibson, J., and Light, P. 1967. Intelligence among university scientists. *Nature* 213:441–43.

Gilbert, C. D. 1977. A study of the interrelationship of factors affecting sixth grade students in respect to mathematics. *School Science and Mathematics* 77:489–94.

Gilbert, C. D., and Cooper, D. 1976. The relationship between teacher/student attitudes and the competency levels of sixth grade students. *School Science and Mathematics* 76:469–76.

Glennon, V. J., and Callahan, L. G. 1968. *A Guide to Current Research: Elementary School Mathematics.* Washington, D.C.: Association for Supervision and Curriculum Development.

Gough, M. F. 1954. Mathemaphobia: Causes and treatments. *Clearing House* 28:290–92.

Graf, R. G., and Riddell, J. C. 1972. Sex differences in problem solving as a function of problem context. *Journal of Educational Research* 65:451–52.

Guay, R. B., and McDaniel, E. D. 1977. The relationship between mathematics achievement and spatial abilities among elementary school children. *Journal for Research in Mathematics Education* 8:211–15.

Guilford, J. P. 1967. *The Nature of Human Intelligence.* New York: McGraw-Hill Book Co.

Guilford, J. P.; Hoepfner, R.; and Peterson, H. 1965. Predicting achievement in ninth-grade mathematics from measures of intellectual-aptitude factors. *Educational and Psychological Measurement* 25:659–82.

Harmon, L. 1971. The childhood and adolescent career plans of college women. *Journal of Vocational Behavior* 1:45–46.

Harrington, L. G. 1960. Attitude toward mathematics and the relationship between such attitude and grade obtained in a freshman mathematics course. Doctoral dissertation, University of Florida. Ann Arbor, Mich.: University Microfilms, No. 60–1901.

Haven, E. W. 1971. Factors associated with the selection of advanced academic mathematics courses by girls in high school. Doctoral dissertation, University of Pennsylvania. Ann Arbor, Mich.: University Microfilms, No. 71–26027.

Helson, R. 1966. Personality of women with imaginative and artistic interests: The role of masculinity, originality, and other characteristics in their creativity. *Journal of Personality* 34:1–25.

_____. 1967a. Personality characteristics and developmental history of creative college women. *Genetic Psychology Monographs* 76:205–56.

_____. 1967b. Sex differences in creative style. *Journal of Personality* 35:214–33.

_____. 1968. Generality of sex differences in creative style. *Journal of Personality* 36:33–48.

Helson, R., and Crutchfield, R. S. 1970. Creative types in mathematics. *Journal of Personality* 38:177–97.

_____. 1970. Mathematicians: The creative researcher and the average Ph.D. *Journal of Consulting and Clinical Psychology* 34:250–7.

_____. 1971. Women mathematicians and the creative personality. *Journal of Consulting and Clinical Psychology* 36:210–20.

Hilton, T. L., and Berglund, G. W. 1974. Sex differences in mathematics achievement—A longitudinal study. *Journal of Educational Psychology* 67:231–37:

Hollingshead, A. B., and Redlich, R. C. 1958. *Social Class and Mental Illness.* New York: Wiley.

Holly, K. A.; Purl, M. C.; Dawson, J. A.; and Michael, W. B. 1973. The relationship of an experimental form of the mathematics self-concept scale to cognitive and noncognitive variables for a sample of seventh-grade pupils in a middle-class southern California community. *Educational and Psychological Measurement* 33:505-8.

Houts, P. S., and Entwisle, D. R. 1968. Academic achievement effort among females: Achievement attitudes and sex-role orientation. *Journal of Counseling Psychology* 15:284-86.

Hoyt, T., and Magoon, T. 1954. A validation study of the Taylor Manifest Anxiety Scale. *Journal of Clinical Psychology* 10:357-61.

Hudson, L. 1960a. Degree class and attainment in scientific research. *British Journal of Psychology* 51:67-73.

_____. 1960b. A differential test of arts-science aptitude. *Nature* 186:413-14.

_____. 1962a. Intelligence, divergence and potential originality. *Nature* 196:601-2.

_____. 1962b. Personality and scientific aptitude. *Nature* 198:913-14.

_____. 1963. The relation of psychological test scores to academic bias. *British Journal of Educational Psychology* 33:120-31.

_____. 1964. Future open scholars. *Nature* 202:834.

_____. 1966. *Contrary Imaginations: A Psychological Study of the Young Student.* New York: Schocken Books.

_____. 1967a. Arts and sciences: The influence of stereotypes on language. *Nature* 186:968-69.

_____. 1967b. The stereotypical scientist. *Nature* 213:228-29.

_____. 1968. *Frames of Mind.* London: Methuen.

_____. 1973. Fertility in the arts and sciences. *Science Studies* 3:305-10.

Hudson, L., and Jacot, B. 1971. Marriage and fertility in academic life. *Nature* 229:531-32.

Hungerman, A. D. 1967. Achievement and attitude of sixth-grade pupils in conventional and contemporary mathematics programs. *Arithmetic Teacher* 14:30-39.

Hunkler, R. 1977. The relationship between a sixth-grade student's ability to predict success in solving computational and statement problems and his mathematics achievement and attitude. *School Science and Mathematics* 77:461-68.

Hunkler, R., and Quast, W. G. 1972. Improving the mathematics attitudes of prospective elementary school teachers. *School Science and Mathematics* 72:709-14.

Husen, T., ed. 1967. *International Study of Achievement in Mathematics: A Comparison of Twelve Countries.* Vols. I & II. New York: John Wiley & Sons.

Husen, T.; Fagerlind, I.; and Liljefors, R. 1974. Sex differences in science achievement and attitudes: A Swedish analysis by grade level. *Comparative Education Review* 18:292-304.

Jacobs, J. E. 1976. Sex-related differences in achievement in mathematics and selected noncognitive factors. Unpublished manuscript. Available through Richmond College, City University of New York, Stuy-vesant Place,Staten Island, New York, NY 10301.

Jarvis, O. T. 1964. Boy-girl ability differences in elementary school arithmetic. *School Science and Mathematics* 64:657-59.

Jones, C. L., and McPherson, A. F. 1973. Fertile imaginations and contrary findings: A comment on subject specialization and sexuality. *Science Studies* 3:389-91.

Josephs, A. P., and Smithers, A. G. 1975. Personality characteristics of syllabus-bound and syllabus-free sixth-formers. *British Journal of Educational Psychology* 45:28-29.

Kaminski, D. M.; Erickson, E. L.; Ross, M.; and Bradfield, L. 1976. Why females don't like mathematics: The effect of parental expectations. Paper presented at American Sociological Association meeting, August 1976, at New York.

Kane, R. B. 1968. Attitudes of prospective elementary school teachers toward mathematics and three other subject areas. *Arithmetic Teacher* 15:169-75.

Keating, D. P., ed. 1976. *Intellectual Talent: Research and Development.* Baltimore: The Johns Hopkins University Press.

Keeves, J. 1973. Differences between the sexes in mathematics and science courses. *International Review of Education* 19:47-62.

Knaupp, J. 1973. Are children's attitudes toward learning arithmetic really important? *School Science and Mathematics* 73:9-15.

Kogan, N., and Pankove, E. 1972. Creative ability over a five-year span. *Child Development* 43:427-42.

Kreinberg, N. 1977. *I'm Madly in Love with Electricity.* Available through the author at Lawrence Hall of Science, Berkeley, CA 94720.

Lambert, P. 1960. Mathematical ability and masculinity. *Arithmetic Teacher* 7:19-21.

Lazarus, M. 1974. Mathophobia: Some personal speculations. *National Elementary Principal* 53:16-22.

_____.1975. Rx for mathophobia. *Saturday Review,* June 28.

Leder, G. C. 1974. Sex differences in mathematics problem appeal as a function of problem context. *Journal of Educational Research* 67:351-53.

Levitt, E. E. 1967. *The Psychology of Anxiety.* Indianapolis: Bobbs-Merrill Co.

Levy, N. 1958. A short form of the Children's Manifest Anxiety Scale. *Child Development* 27:153-54.

Lyda, W. J., and Morse, E. C. 1963. Attitudes, teaching methods, and arithmetic achievement. *Arithmetic Teacher* 10:136-38.

Lynn, D. B. 1972. Determinants of intellectual growth in women. *School Review* 80:241-60.

MacArthur, R. 1967. Sex differences in field dependence for the Eskimo. *International Journal of Psychology* 2:139-40.

MacKay, C. K., and Cameron, M. G. 1968. Cognitive bias in Scottish first-year science and arts undergraduates. *British Journal of Educational Psychology* 38:315–18.

MacKinnon, D. W. 1962. The nature and nurture of creative talent. *American Psychologist* 17:484–95.

_____. 1965. Personality and the realization of creative potential. *American Psychologist* 20:273–81.

McCallon, E. L., and Brown, J. D. 1971. Semantic differential instrument for measuring attitude toward mathematics. *Journal of Experimental Education* 39:69–72.

McCarthy, J. L., and Wolfle, D. 1975. Doctorates granted to women and minority group members. *Science* 189:856–59.

McNarry, L. R., and O'Farrell, S. 1971. Students reveal negative attitudes toward technology. *Science* 172:1060–61.

Maccoby, E. E., and Jacklin, C. N. 1974. *Psychology of Sex Differences.* Stanford: Stanford University Press.

Mager, R. F. 1968. *Developing attitude toward learning.* Palo Alto, Calif.: Fearon Publishers.

Maier, N. R. F. and Casselman, G. C. 1971. Problem-solving ability as a factor in selection of major in college study: Comparison of the processes of "idea-getting" and "making essential distinctions" in males and females. *Psychological Reports* 28:503–14.

Mandler, G., and Sarason, S. 1952. A study of anxiety and learning. *Journal of Abnormal and Social Psychology* 47:166–73.

Mead, M., and Metraux, R., 1957. Image of the scientist among high school students. *Science* 126:384–90.

Milton, G. A. 1957. The effects of sex-role identification upon problem-solving skill. *Journal of Abnormal and Social Psychology* 55:208–12.

Mitias, R. G. E. 1970. Concepts of science and scientists among college students. *Journal of Research in Science Teaching* 7:135–40.

Mullis, I. V. S. 1975. *Educational Achievement and Sex Discrimination.* Denver, Colorado: National Assessment of Educational Progress.

Muscio, R. D. 1962. Factors related to quantitative understanding in the sixth grade. *Arithmetic Teacher* 9:258–62.

Mokros, J. R., and Koff, E. 1978. Sex stereotyping of children's success in mathematics and reading. Psychological Reports 42:1287–93.

National Assessment of Educational Progress. August 1979. *Mathematical Knowledge and Skills: Selected Results from the Second Assessment of Mathematics.* Report No. 09-MA-02. Denver, Colo.: Education Commission of the States.

National Center for Education Statistics, U.S. Department of Health, Education, and Welfare. August 1979. *The Relationship between Participation in Mathematics at the High School Level and Entry into Quantitative Fields: Results from the National Longitudinal Study.* Washington, D.C.

Naylor, F. D.; and Gaudry, E. 1973. The relationship of adjustment, anxiety, and intelligence to mathematics performance. *Journal of Educational Research* 66:413-17.

Neale, D. C. 1969. The role of attitudes in learning mathematics. *Arithmetic Teacher* 16:631-40.

Nevin, M. 1973. Sex differences in participation rates in mathematics and science at Irish schools and universities. *International Review of Education* 19:89-91.

Nicholls, J. G. 1972. Creativity in the person who will never produce anything original and useful: The concept of creativity as a normally distributed trait. *American Psychologist* 27:717-27.

Ohlson, E. L., and Mein, L. 1977. The difference in level of anxiety in undergraduate mathematics and nonmathematics majors. *Journal for Research in Mathematics Education* 8:211-15.

Parsley, K. M.; Powell, M.; and O'Connor, H. A. 1964. Further investigation of sex differences in achievement of under-average, and over-achieving students within five IQ groups in grades four through eight. *Journal of Educational Research* 57:268-70.

Parsley, K. M.; Powell, M.; O'Connor, H. A.; and Deutsch, M. 1963. Are there really sex differences in achievement? *Journal of Educational Research* 57:210-12.

Parsons, J. 1979. Developmental shifts in expectancies, utility values, and attributions for performance in mathematics. Paper presented at the Society for Research in Child Development meetings, March 18, in San Francisco, California. For copies, write Dr. Jacqueline E. Parsons, Psychology Department, University of Michigan, 3433 Mason Hall, Ann Arbor, MI 48109.

Perl, T. 1978 *Math Equals: Biographies of Women Mathematicians and Related Activities.* Reading, Pa.: Addison-Wesley.

Perucci, C. C. 1970. Minority status and the pursuit of professional careers: Women in science and engineering. *Social Forces* 49:245-59.

Poffenberger, T., and Norton, D. A. 1956. Factors determining attitudes toward arithmetic and mathematics. *Arithmetic Teacher* 3:113-16.

Povey, R. M. 1970. Arts/Science differences: Their relationship to curriculum specialization. *British Journal of Psychology* 61:55-64.

Psathas, G. 1968. Toward a theory of occupational choice for women. *Sociology and Social Research* 52:253-68.

Rand, L. 1968. Masculinity or femininity? Differentiating career-oriented and homemaking-oriented college freshman women. *Journal of Counseling Psychology* 15:284-86.

Reys, R. E., and Delon, F. G. 1968. Attitudes of prospective elementary school teachers towards arithmetic. *Arithmetic Teacher* 15:363-66.

Rezler, A. G. 1967. Characteristics of high school girls choosing traditional or pioneer vocations. *Personnel and Guidance Journal* 45:659-65.

Richards, P. N., and Bolton, N. 1971. Type of mathematics teaching, mathematical ability, and divergent thinking in junior school children. *British Journal of Educational Psychology* 42:32-37.

Richardson, F. C., and Suinn, R. M. 1972. The Mathematics Anxiety Rating Scale: Psychometric data. *Journal of Counseling Psychology* 19:551-54.

_____. 1973. A comparison of traditional systematic desensitization, accelerated massed desensitization, and anxiety management training in the treatment of mathematics anxiety. *Behavior Therapy* 4:212-18.

Roberts, F. 1969. Attitudes of college freshmen towards mathematics. *Mathematics Teacher* 62:25-27.

Romberg, T. A. 1969. Current research in mathematics education. *Review of Educational Research* 39:473-91.

Rosenkrantz, P.; Vogel, S.; Bee, H.; Broverman, I.; and Broverman, D. 1968. Sex-role stereotypes and self-concepts in college students. *Journal of Consulting and Clinical Psychology* 32:356-63.

Rosenthal, R., and Jacobson, L. 1968. *Pygmalion in the Classroom*. New York: Holt, Rinehart, and Winston, Inc.

Rossi, A. S. 1965. Women in science: Why so few? *Science* 148:1196-1202.

Rump, E. E., and Dunn, M. 1971. Extensions to the study of science students' divergent thinking ability. *Nature* 229:349-50.

Sarason, S. B.; Lighthall, F. F.; Davidson, K. S.; Waite, R. R.; and Ruebush, B. K. 1960. *Anxiety in Elementary School Children*. New York: John Wiley & Sons.

Sarason, I. G., and Winkle, G. H. 1966. Individual differences among subjects and experimenters and subjects' self-descriptions. *Journal of Personality and Social Psychology* 3:448-57.

Schwandt, A. K.; Kreinberg, N.; and Downie, D. May 1979. Use of EQUALS to promote participation of women in mathematics. Mimeographed. Berkeley, Calif.: EQUALS Institute, University of California, Berkeley.

Sells, L. 1974. Fact sheet on women in higher education. Berkeley Chapter of California Women in Higher Education. Available through Women's Center, Building T-9, University of California, Berkeley, CA 94720.

_____.1975. Sex and discipline differences in doctoral attrition. Doctoral dissertation, University of California, Berkeley. Ann Arbor, Mich. University Microfilms, No. 76-15363. 1976.

Sepie, A. C., and Keeling, B. 1978. The relationship between types of anxiety and under-achievement in mathematics. *Journal of Educational Research* 72:15-19.

Sharples, D. 1969. Children's attitudes towards junior school activities. *British Journal of Educational Psychology* 39:72-77.

Shedletsky, R., and Endler, N. S. 1974. Anxiety: The state-trait model and the interaction model. *Journal of Personality* 42:511-27.

Sheehan, T. J. 1968. Patterns of sex differences in learning mathematical problem solving. *Journal of Experimental Education* 36:84-87.

Shepps, F. P., and Shepps, R. R. 1971. Relationship of study habits and school attitudes to achievement in mathematics and reading. *Journal of Educational Research* 65:71–73.

Sherman, J. 1967. Problems of sex differences in space perception and aspects of intellectual functioning. *Psychological Review* 74:290–99.

Sherman, J. 1976. Effects of biological factors on sex-related differences in mathematics achievement. Paper prepared for National Institute of Education.

_____. 1978. *Sex-Related Cognitive Differences: An Essay on Theory and Evidence*. Springfield: Charles C. Thomas.

Sherman, J., and Fennema, E. 1977. The study of mathematics by high school girls and boys: variables. *American Educational Research Journal* 14:159–68.

Silverman, S. M. 1974. Parental loss and scientists. *Science Studies* 4:259–64.

Smith, F. 1964. Prospective teachers' attitudes toward arithmetic. *Arithmetic Teacher* 11:474–77.

Smithers, A. 1969. Occupational values of students. *Nature* 222:725–26.

Smithers, A., and Child, D. 1974. Convergers and divergers: Different forms of neuroticism? *British Journal of Educational Psychology* 44:304–6.

Speilberger, C. D. 1966. *Anxiety and Behavior*. New York: Academic Press.

Speilberger, C. D.; Gorsuch, R. L.; and Lushene, R. E. 1970. *STAI Manual for the State-Trait Anxiety Inventory*. Palo Alto, Calif.: Consulting Psychologists Press.

Stafford, R. E. 1972. Hereditary and environmental components of quantitative reasoning. *Review of Educational Research* 42:183–201.

Stallings, J., and Robertson, A. 1979. Factors influencing women's decisions to enroll in advanced mathematics courses. A final report prepared for the National Institute of Education under Grant No. NIE-G-78-0024. Menlo Park, Calif.: SRI International.

Stanley, J. C.; Keating, D. P.; and Fox, L. H., eds. 1974. *Mathematical Talent: Discovery, Description and Development*. Baltimore: Johns Hopkins University Press.

Steele, L., and Wise, L. L. 1979. Origins of sex differences in high school mathematics achievement and participation. Paper presented at the American Educational Research Association meetings, April 1979, in San Francisco, California.

Stein, A. H., and Smithells, J. 1969. Age and sex differences in children's sex-role standards about achievement. *Developmental Psychology* 1:252–59.

Stright, V. M. 1960. A study of the attitudes toward arithmetic of students and teachers in the third, fourth, and sixth grades. *Arithmetic Teacher* 7:280–86.

Suinn, R. M. 1969. The STABS, a measure of test anxiety for behavior therapy: Normative data. *Behavior Research and Therapy* 7:335–39.

Suinn, R. M.; Edie, C. A.; Nicoletti, J.; and Spinelli, P. R. 1972. The MARS, a measure of mathematics anxiety: Psychometric data. *Journal of Clinical Psychology* 28:373–75.

Taylor, J. A. 1953. A personality scale of manifest anxiety. *Journal of Abnormal and Social Psychology* 48:285–90.

Thurlow, V. 1965. Mathematical understanding of seventh- and eighth-grade pupils, 1948 and 1963. *Arithmetic Teacher* 12:43–44.

Tobias, S. 1976. Math anxiety. *Ms.* 4:56–59.

_____. 1978. *Overcoming Math Anxiety.* New York: W. W. Norton & Co.

Todd, R. M. 1966. A mathematics course for elementary teachers: Does it improve understanding and attitude? *Arithmetic Teacher* 13:198–202.

Torrance, E. P. 1962. *Guiding Creative Talent.* Englewood Cliffs, N.J.: Prentice-Hall.

_____. 1965. *Rewarding Creative Behavior.* Englewood Cliffs, N.J.: Prentice-Hall.

U.S. Bureau of the Census. 1977. *Current Population Reports,* Series P-25, No. 704. Washington, D.C.: U.S. Government Printing Office.

U.S. Department of Labor. 1976. *Occupational Outlook Handbook, 1976–77 Edition.* Washington, D.C.: U.S. Government Printing Office.

_____. 1977. Single men and married women show unusually large labor force gains. *News,* September 14:77.

Unkel, E. 1966. A study of the interaction of socioeconomic groups and sex factors with the discrepancy between anticipated achievement and actual achievement in elementary school mathematics. *Arithmetic Teacher* 13:662–70.

Vernon, P. E. 1971. Effects of administration and scoring on divergent thinking tests. *British Journal of Educational Psychology* 41:245–57.

Very, P. S. 1967. Differential factor structures in mathematical ability. *Genetic Psychology Monographs* 75:169–207.

Wainer, H. 1976. Estimating coefficients in linear models: It don't make no nevermind. *Psychological Bulletin* 83:213–17.

Wallach, M. A., and Kogan, N. 1965. *Modes of Thinking in Young Children.* New York: Holt, Rinehart, & Winston.

Ward, W. C. 1968. Creativity in young children. *Child Development* 38:737–54.

_____. 1969. Rate and uniqueness in children's creative responding. *Child Development* 40:869–78.

Ward, W. C., and Kogan, N. 1972. Incentive effects in children's creativity. *Child Development* 43:669–76.

White, M. S. 1970. Psychological and social barriers to women in science. *Science* 170:413–16.

Wilson, G. M. 1961. Why do pupils avoid mathematics in high school? *Arithmetic Teacher* 8:168–71.

Wilson, J. W. 1972. *Patterns of Mathematics Achievement in Grade 11: Z Population.* National longitudinal study of mathematical abilities, No. 17. Palo Alto, Calif.: Stanford University Press.

Wise, L. L. 1978. The role of mathematics in women's career development. Paper presented at the American Psychological Association meetings, August 1978, in Toronto, Canada.

————. 1979. Long-term consequences of sex differences in high school mathematics education. Paper presented at the American Educational Research Association meetings, April 1979, in San Francisco, California.

Women's Bureau. 1975. *1975 Handbook on Women Workers.* Washington, D.C.: U.S. Department of Labor.

Woodward, W. R. 1974. Scientific genius and loss of a parent. *Science Studies* 4:265–77.

Wozencraft, M. 1963. Are boys better than girls in arithmetic? *Arithmetic Teacher* 10:486–90.

Zuckerman, M. 1960. The development of an affect adjective checklist for the measurement of anxiety. *Journal of Consulting Psychology* 24:457–62.

Appendix A

Tenth-Grade Questionnaire

Number _____

Name_____ Sex (circle one) M F   Age in Years_____

Grade in School_____ Name of School_____

1. Ethnic background (check one):

Caucasian (not of Hispanic origin) _____   Hispanic_____

Black (not of Hispanic origin) _____   Asian _____

Native American _____   Other (please specify) _____

2. Father's occupation_____

3. Father's education (check highest level completed):

eighth grade or less _____

some high school _____

high school diploma _____

some college or two-year college degree _____

four-year college degree _____

study beyond college _____

4. Mother's occupation _____

5. Mother's education (check highest level completed):

eighth grade or less _____

some high school _____

high school diploma _____

some college or two-year college degree _____

four-year college degree _____

study beyond college _____

6. How many older brothers do you have?__ __

How many younger brothers do you have?_____

How many older sisters do you have?_____

How many younger sisters do you have?_____

7. Please write down the number of years you have studied and/or intend to study each of the following subjects while in high school.

    Astronomy ____   Mathematics ____

    Chemistry ____   Physics ____

    A Foreign Language ____   Social Studies/History

8. Suppose you had to take a course in each of the following subjects. Rate how much you think you would enjoy each subject. Use a scale from 1 to 5 where 1 = I would <u>dislike</u> it very much; 2 = I would dislike it somewhat; 3 = I am indifferent; 4 = I would like it somewhat; 5 = I would <u>like</u> it very much.

    Biology ____   Languages ____

    Chemistry ____   Mathematics ____

    English ____   Physics ____

    History ____

9. Suppose you had to take a course in each of the following subjects. For each subject predict the grade you would get from F to A.

    Biology ____   Languages ____

    Chemistry ____   Mathematics ____

    English ____   Physics ____

    History ____

10. Do you plan to go to college?   Yes ____   No ____

11. What are your career plans?   (Please be specific)

    First choice _____

    Second choice _____

    Third choice _____

On the following pages are some statements. There are no correct answers for these statements. They have been written so that you may agree or disagree with the idea expressed. Suppose the statement is:

Example 1. I like mathematics............................ SA A U D SD

As you read the statement, you will know whether you agree or disagree. If you strongly agree, circle the letters SA (Strongly Agree) following the statement. If you agree, but have some reservations, circle A (Agree). If you disagree with the idea, indicate the extent to which you disagree by circling D (Disagree) or SD (Strongly Disagree). But if you neither agree nor disagree, that is, you are not certain, circle U (Uncertain). Now circle your response to Example 1, and do the same for Example 2 below.

Example 2. English is very interesting to me............ SA A U D SD

Do not spend much time with any statement, but be sure to answer every statement. Work fast, but carefully.

1.   Anyone can learn mathematics............................. SA A U D SD

2.   I am sure I could do advanced work in English............. SA A U D SD

3.   My parents expect me to do well in math.................. SA A U D SD

4.   I enjoy reading books for my homework in English......... SA A U D SD

5.   There is not much room for my own ideas in math.......... SA A U D SD

6.   Females write more beautiful poems than males............ SA A U D SD

7.   In English there is a lot of learning by heart........... SA A U D SD

8.   My teachers do not expect me to do well in English....... SA A U D SD

9.   Taking a test in English class doesn't bother me......... SA A U D SD

10. Knowing mathematics will help me earn a living........... SA A U D SD

11. I would trust a woman just as much as I would trust
     a man to figure out important calculations............... SA A U D SD

12. Studying English now will help me earn a living later on.. SA A U D SD

13. My friends don't think of me as a math whiz.............. SA A U D SD

14. I get scared when I open my math book and see a page
     full or problems... SA A U D SD

15. Math has been my worst subject........................... SA A U D SD

16. Boys who enjoy studying English are a bit peculiar....... SA A U D SD

SA--Strongly Agree; A--Agree; U--Uncertain; D--Disagree; SD--Strongly Disagree

17. English is a fairly easy subject.......................... SA A U D SD

18. My friends believe that I am good in English............. SA A U D SD

19. The challenge of math problems does not appeal to me...... SA A U D SD

20. I put off studying for a math test as long as I can....... SA A U D SD

21. It's hard to believe a female could be a genius in
    mathematics... SA A U D SD

22. Mathematics is a worthwhile and necessary subject......... SA A U D SD

23. My parents do not encourage me in my study of math........ SA A U D SD

24. I do as little work in English as possible................ SA A U D SD

25. I look forward to having to report to the class about
    a book I read... SA A U D SD

26. Males are as good as females at writing poetry............ SA A U D SD

27. Math is fairly easy....................................... SA A U D SD

28. Outside of journalism and writing books, there is little
    need for writing skills in most jobs...................... SA A U D SD

29. My teachers are discouraging about my ability in English.. SA A U D SD

30. I get a sinking feeling when I think of trying to write
    a composition on a hard topic............................. SA A U D SD

31. Mathematics is dull and monotonous....................... SA A U D SD

32. I would have more faith in the answer for a math problem
    solved by a man than a woman.............................. SA A U D SD

33. Anyone can learn to write good compositions............... SA A U D SD

34. My friends support me in my study of math................. SA A U D SD

35. Calculating percentages is fun............................ SA A U D SD

36. I expect to have little use for math when I get out of
    school.. SA A U D SD

37. Boys can do just as well as girls in English.............. SA A U D SD

SA--Strongly Agree; A--Agree; U--Uncertain; D--Disagree; SD--Strongly Disagree

38. I don't think I could do advanced mathematics.............. SA A U D SD

39. My parents expect me to do well in English................. SA A U D SD

40. My own ideas can be used in English class.................. SA A U D SD

41. I feel nervous during math tests.......................... SA A U D SD

42. Boys tend to be better at math than girls................. SA A U D SD

43. When a math problem arises that I can't immediately solve,
    I stick with it until I have the solution................. SA A U D SD

44. My teachers are discouraging about my ability in math...... SA A U D SD

45. I study English because I know how useful it is............ SA A U D SD

46. I look forward to taking more English courses.............. SA A U D SD

47. Women are better than men at expressing their feelings
    on paper.. SA A U D SD

48. It is important to know math in order to get a good job.... SA A U D SD

49. My friends don't think of me as an English whiz........... SA A U D SD

50. I don't think I could do advanced work in English.......... SA A U D SD

51. I am not worried by assignments which ask me to write
    a poem.. SA A U D SD

52. Women are certainly logical enough to do well in
    mathematics... SA A U D SD

53. I am sure that I can learn mathematics.................... SA A U D SD

54. The work for English class can be exciting................ SA A U D SD

55. My teachers encourage me in my study of math.............. SA A U D SD

56. I do not like being watched as I do a set of division
    problems.. SA A U D SD

57. Mathematics can be exciting............................... SA A U D SD

58. When a man has to write a composition about feelings, it
    is appropriate to ask a woman to help..................... SA A U D SD

APPENDIX A

SA--Strongly Agree; A--Agree; U--Uncertain; D--Disagree; SD--Strongly Disagree

59. Math will not be important to me in my life's work........ SA A U D SD

60. My friends support me in my study of English.............. SA A U D SD

61. It is important to be able to write well in order to get
    a good job.. SA A U D SD

62. I usually have been at ease during math tests............. SA A U D SD

63. Males are not naturally better than females in
    mathematics.. SA A U D SD

64. Very few people can learn mathematics.................... SA A U D SD

65. My teachers do not expect me to do well in mathematics.... SA A U D SD

66. English has been my worst subject........................ SA A U D SD

67. I often clam up when asked to explain the meaning of
    a poem... SA A U D SD

68. Females are not naturally better than males in English.... SA A U D SD

69. I am challenged by math problems I can't understand
    immediately.. SA A U D SD

70. My parents do not encourage me in my study of English..... SA A U D SD

71. The work assigned in English is dull..................... SA A U D SD

72. I do not like having to write compositions in class....... SA A U D SD

73. Girls can do just as well as boys in mathematics.......... SA A U D SD

74. Math is not useful for the problems of everyday life...... SA A U D SD

75. My parents do not think that I can do well in math........ SA A U D SD

76. English is a worthwhile and necessary subject to study.... SA A U D SD

77. Mathematics tends to be difficult........................ SA A U D SD

78. Men are certainly sensitive enough to do well in English.. SA A U D SD

79. It wouldn't bother me at all to take more math courses.... SA A U D SD

80. Math involves a lot of learning by heart.................. SA A U D SD

SA--Strongly Agree; A--Agree; U--Uncertain; D--Disagree; SD--Strongly Disagree

81. My parents encourage me in my study of English............. SA A U D SD

82. English is a difficult subject............................. SA A U D SD

83. I get a sinking feeling when I think of trying hard
     math problems... SA A U D SD

84. When a woman has to solve a math problem, it is
     appropriate to ask a man for help......................... SA A U D SD

85. Taking mathematics is a waste of time..................... SA A U D SD

86. English class gives you the chance to think things
     out for yourself.. SA A U D SD

87. My friends believe that I am good in math................. SA A U D SD

88. It bothers me to have to take English courses all the way
     through high school....................................... SA A U D SD

89. I am sure I could do advanced work in mathematics......... SA A U D SD

90. Girls tend to be better at English than boys.............. SA A U D SD

91. It is not necessary for me to study English............... SA A U D SD

92. My friends don't support me in my study of English........ SA A U D SD

93. I get scared while waiting for the teacher to return
     a composition I wrote..................................... SA A U D SD

94. My own ideas can be used in math.......................... SA A U D SD

95. Girls who enjoy studying math are a bit peculiar.......... SA A U D SD

96. I study math because I know how useful it is.............. SA A U D SD

97. My parents encourage me in my study of mathematics........ SA A U D SD

98. Very few people can learn to write good compositions...... SA A U D SD

99. I enjoy being given a set of addition problems to solve.... SA A U D SD

100. I would trust a man just a much as I would trust a woman to
     write a report of a group activity....................... SA A U D SD

101. I'm not the type to do well in math...................... SA A U D SD

SA--Strongly Agree; A--Agree; U--Uncertain; D--Disagree; SD--Strongly Disagree

102. My teachers encourage me in my study of English............ SA A U D SD

103. There is not much room for my own ideas in English......... SA A U D SD

104. I usually enjoy preparing for a math test.................. SA A U D SD

105. Females are as good as males at working problems about
     fractions... SA A U D SD

106. Math gives me the chance to think things out for myself.... SA A U D SD

107. My teachers think I can do well in math.................... SA A U D SD

108. Taking English is a waste of time.......................... SA A U D SD

109. I get scared when my homework involves correctly
     punctuating sentences....................................... SA A U D SD

110. I would expect a male writer to be somewhat feminine....... SA A U D SD

111. Outside of science and engineering, there is little need
     for mathematics in most jobs............................... SA A U D SD

112. I can get good grades in English........................... SA A U D SD

113. My parents do not think that I can do well in English...... SA A U D SD

114. English doesn't scare me at all............................ SA A U D SD

115. I can get good grades in mathematics....................... SA A U D SD

116. Studying mathematics is just as appropriate for women
     as for men.. SA A U D SD

117. I do as little work in math as possible.................... SA A U D SD

118. My friends don't support me in my study of math............ SA A U D SD

119. Once I start working on an English composition, I find
     it hard to stop.. SA A U D SD

120. Math doesn't worry me at all............................... SA A U D SD

121. Studying English is just as appropriate for men as for
     women... SA A U D SD

122. I will use math in many ways as an adult................... SA A U D SD

123. My teachers think I can do well in English................. SA A U D SD

SA--Strongly Agree; A--Agree; U--Uncertain; D--Disagree; SD--Strongly Disagree

124. My study of English will not help me earn a living........ SA A U D SD

125. I feel uneasy when I realize I must take a certain number
     of math classes to graduate from high school.............. SA A U D SD

126. I would expect a woman mathematician to be a masculine
     type of person.. SA A U D SD

APPENDIX A

Please answer each of the following questions in as much detail as possible:

1. Think about your work in mathematics. Think of a lesson or a unit that you really liked. It could be from this year's work or a year in the past. What kind of problems were you working on?

   Why did you like it so much?

2. Now think about a lesson or unit in math that you really disliked. What kind of problems were you working on?

   What made it so unpleasant?

3. If you were teaching math, what is the first thing you would change from the way it is done now? Remember that you want students to enjoy learning math and to be successful at it.

4. Think about your work in reading, writing, spelling, and grammar. Think about an assignment or a lesson you really enjoyed in these subjects and describe the lesson assigned.

   Why did you like it so much?

5. Now think about a lesson in English that you really disliked. Describe the lesson.

   What made it so unpleasant?

6. If you were teaching English, what is the first thing you would change from the way it is done now? Remember that you want your students to enjoy learning English and to be successful at it.

Number _____

Name _____

The purpose of the next set of scales is to find out the impression you have of people with different occupations. At the top of the next few pages is a description of an individual's occupation. Below is a series of descriptive scales. Please rate each occupation on each scale as follows: If you feel an individual with the specified occupational role is very like one end of the scale, circle the number at that extreme of the scale:

reckless (1) 2 3 4 5 6 7 cautious

If this person is quite like one end of the scale (but not extremely), circle a "2" or a "6", whichever is near the appropriate end of the scale:

radical 1 2 3 4 5 (6) 7 conservative

And so on. The direction toward which you check, of course, depends upon which of the two ends of the scale seems most characteristic of the type of person described. Please reserve the middle number, "4", for neutral judgments, that is, circle "4" if both sides of the scale are equally associated with the person described, or if the traits described are completely irrelevant.

Work fairly quickly through these scales; it is your first impression that we want. However, make sure you mark every scale and that you circle only one number for each item.

This person is a writer. Sometimes the job demands teaching literature and/or poetry courses in a university; lots of time is spent writing articles, poems, or books.

1. Very concerned with financial success   1   2   3   4   5   6   7   not all concerned with financial success

2. not at all warm and affectionate   1   2   3   4   5   6   7   very warm and affectionate

3. radical   1   2   3   4   5   6   7   conservative

4. very assertive   1   2   3   4   5   6   7   not at all assertive

5. very creative   1   2   3   4   5   6   7   not at all creative

6. individualist   1   2   3   4   5   6   7   conformist

7. calm   1   2   3   4   5   6   7   excitable

8. low opportunity for advancement   1   2   3   4   5   6   7   high opportunity for advancement

9. sensitive   1   2   3   4   5   6   7   insensitive

10. reckless   1   2   3   4   5   6   7   cautious

11. very independent   1   2   3   4   5   6   7   not at all independent

12. very gentle   1   2   3   4   5   6   7   not at all gentle

13. not at all interested in art   1   2   3   4   5   6   7   very interested in art

14. not at all intuitive   1   2   3   4   5   6   7   very intuitive

15. irresponsible   1   2   3   4   5   6   7   responsible

16. very powerful over others   1   2   3   4   5   6   7   not at all powerful over others

17. wise   1   2   3   4   5   6   7   foolish

18. rational   1   2   3   4   5   6   7   irrational

19. not at all competitive   1   2   3   4   5   6   7   very competitive

20. sociable   1   2   3   4   5   6   7   retiring

21. changeable   1   2   3   4   5   6   7   stable

This individual is a mathematician. Sometimes the job entails teaching mathematics in a university; lots of time is spent thinking about mathematics and proving new theorems.

1. very concerned with financial success   1 2 3 4 5 6 7   not all concerned with financial success

2. not at all warm and affectionate   1 2 3 4 5 6 7   very warm and affectionate

3. radical   1 2 3 4 5 6 7   conservative

4. very assertive   1 2 3 4 5 6 7   not at all assertive

5. very creative   1 2 3 4 5 6 7   not at all creative

6. individualist   1 2 3 4 5 6 7   conformist

7. calm   1 2 3 4 5 6 7   excitable

8. low opportunity for advancement   1 2 3 4 5 6 7   high opportunity for advancement

9. sensitive   1 2 3 4 5 6 7   insensitive

10. reckless   1 2 3 4 5 6 7   cautious

11. very independent   1 2 3 4 5 6 7   not at all independent

12. very gentle   1 2 3 4 5 6 7   not at all gentle

13. not at all interested in art   1 2 3 4 5 6 7   very interested in art

14. not at all intuitive   1 2 3 4 5 6 7   very intuitive

15. irresponsible   1 2 3 4 5 6 7   responsible

16. very powerful over others   1 2 3 4 5 6 7   not at all powerful over others

17. wise   1 2 3 4 5 6 7   foolish

18. rational   1 2 3 4 5 6 7   irrational

19. not at all competitive   1 2 3 4 5 6 7   very competitive

20. sociable   1 2 3 4 5 6 7   retiring

21. changeable   1 2 3 4 5 6 7   stable

Now, please rate yourself on the following scales.

1.	very concerned with financial success	1	2	3	4	5	6	7	not all concerned with financial success
2.	not at all warm and affectionate	1	2	3	4	5	6	7	very warm and affectionate
3.	radical	1	2	3	4	5	6	7	conservative
4.	very assertive	1	2	3	4	5	6	7	not at all assertive
5.	very creative	1	2	3	4	5	6	7	not at all creative
6.	individualist	1	2	3	4	5	6	7	conformist
7.	calm	1	2	3	4	5	6	7	excitable
8.	low opportunity for advancement	1	2	3	4	5	6	7	high opportunity for advancement
9.	sensitive	1	2	3	4	5	6	7	insensitive
10.	reckless	1	2	3	4	5	6	7	cautious
11.	very independent	1	2	3	4	5	6	7	not at all independent
12.	very gentle	1	2	3	4	5	6	7	not at all gentle
13.	not at all interested in art	1	2	3	4	5	6	7	very interested in art
14.	not at all intuitive	1	2	3	4	5	6	7	very intuitive
15.	irresponsible	1	2	3	4	5	6	7	responsible
16.	very powerful over others	1	2	3	4	5	6	7	not at all powerful over others
17.	wise	1	2	3	4	5	6	7	foolish
18.	rational	1	2	3	4	5	6	7	irrational
19.	not at all competitive	1	2	3	4	5	6	7	very competitive
20.	sociable	1	2	3	4	5	6	7	retiring
21.	changeable	1	2	3	4	5	6	7	stable

Appendix B
Additional Tables

TABLE 1.
Student Means on the Easy/Difficult Scale

ITEM MEANS[a]

ITEMS	Grade 6 M	Grade 6 F	Grade 7 M	Grade 7 F	Grade 8 M	Grade 8 F	Grade 9 M	Grade 9 F	Grade 10 M	Grade 10 F	Grade 11 M	Grade 11 F	Grade 12 M	Grade 12 F
Anyone can learn mathematics	3.41	3.38	3.50	3.43	3.77	3.61	3.57	3.50	3.65	3.49	3.78	3.57	3.27	3.05
Very few people can learn mathematics	3.93	4.03	4.21	4.20	4.25	4.33	4.23	4.05	4.17	4.16	4.20	4.08	3.99	4.07
Math is fairly easy	3.65	3.55	3.41	3.30	3.54	3.18	3.09	3.03	3.26	3.15	3.33	2.95	2.96	2.80
Mathematics tends to be difficult	2.94	2.73	2.73	2.48	2.71	2.60	2.61	2.18	2.53	2.29	2.57	2.40	2.44	2.30
I am sure that I can learn mathematics	4.49	4.34	4.30	4.06	4.44	4.31	4.21	3.99	4.14	4.04	4.25	3.96	4.07	3.91
I'm not the type to do well in math	3.73	3.46	3.67	3.36	3.75	3.47	3.58	3.07	3.51	3.23	3.58	3.19	3.33	3.12
I can get good grades in mathematics	4.02	3.85	3.91	3.66	4.00	3.80	3.76	3.42	3.86	3.61	3.87	3.40	3.67	3.42
Math has been my worst subject	3.88	3.91	3.93	3.60	3.96	3.69	3.78	3.30	3.66	3.52	3.76	3.35	3.58	3.28
I am sure I could do advanced work in mathematics	3.50	3.27	3.57	3.03	3.55	3.23	3.46	2.99	3.40	2.89	3.37	2.83	3.46	2.91
I don't think I could do advanced mathematics	2.67	2.93	3.58	3.00	3.53	3.28	3.05	2.85	3.27	2.77	3.44	2.77	3.49	3.16

[a]Higher score reflects more positive attitude about mathematics.

TABLE 2.
Student Means on the Enjoyable/Anxiety-Provoking Scale

ITEMS	Grade 6		Grade 7		Grade 8		Grade 9		Grade 10		Grade 11		Grade 12	
	M	F	M	F	M	F	M	F	M	F	M	F	M	F
I enjoy being given a set of addition problems to solve	3.72	3.68	3.46	3.52	3.33	3.35	3.18	3.17	2.99	3.25	3.02	3.16	2.87	3.20
I get scared when I open my math book and see a page full of problems	3.67	3.43	3.70	3.52	3.94	3.71	3.91	3.55	3.92	3.50	3.93	3.32	3.76	3.54
I usually enjoy preparing for a math test	3.10	2.66	2.72	2.34	2.70	2.33	2.49	2.18	2.48	2.10	2.45	2.11	2.15	2.08
I put off studying for a math test as long as I can	3.61	3.80	3.53	3.70	3.42	3.72	3.45	3.29	3.26	3.24	3.13	3.16	3.13	3.33
I usually have been at ease during math tests	3.46	2.98	3.37	2.80	3.54	2.99	3.25	2.84	3.35	2.81	3.23	2.87	2.94	2.71
I feel nervous during math tests	3.23	2.62	3.08	2.47	3.41	2.66	3.36	2.63	3.30	2.69	3.19	2.73	3.04	2.64
Calculating percentages is fun	3.44	3.06	2.94	2.91	2.98	2.87	2.86	2.50	2.86	2.37	2.55	2.36	2.48	2.36
I do not like being watched as I do a set of division problems	2.29	2.36	2.39	2.11	2.62	2.28	2.57	2.08	2.60	2.28	2.70	2.38	2.80	2.42
Math doesn't worry me at all	3.59	3.32	3.39	2.90	3.54	2.99	3.26	2.73	3.29	2.69	3.01	2.65	3.09	2.57
I get a sinking feeling when I think of trying hard math problems	3.37	3.05	3.31	2.96	3.40	3.13	3.09	3.01	3.26	3.05	3.35	2.86	3.16	2.96
It wouldn't bother me at all to take more math courses	3.34	3.38	3.25	2.97	3.31	3.10	3.21	2.88	2.97	2.82	3.06	2.72	2.81	2.77
I feel uneasy when I realize I must take a certain number of math classes to graduate from high school	2.97	3.01	3.03	3.01	3.27	3.16	3.59	3.32	3.62	3.48	3.72	3.32	3.64	3.72

ITEM MEANS[a]

[a]Higher score reflects more positive attitude about mathematics.

TABLE 3.
Student Means on the Creative/Dull Scale

ITEM MEANS[a]

ITEMS	Grade 6 M	Grade 6 F	Grade 7 M	Grade 7 F	Grade 8 M	Grade 8 F	Grade 9 M	Grade 9 F	Grade 10 M	Grade 10 F	Grade 11 M	Grade 11 F	Grade 12 M	Grade 12 F
Mathematics can be exciting	3.76	3.71	3.59	3.44	3.55	3.43	3.38	3.03	3.24	3.09	3.59	3.13	3.44	3.16
Mathematics is dull and monotonous	3.66	3.79	3.49	3.56	3.55	3.47	3.54	3.21	3.17	3.20	3.40	3.22	3.15	3.16
Math gives me the chance to think things out for myself	3.65	3.26	3.54	3.33	3.45	3.24	3.20	2.88	3.28	2.92	3.27	2.96	2.83	2.66
Math involves a lot of learning by heart	2.17	2.37	2.23	2.70	2.59	2.69	2.95	2.68	2.78	2.75	2.80	2.64	2.67	2.45
My own ideas can be used in math	3.10	3.12	3.03	2.99	3.08	2.90	2.92	2.56	2.72	2.54	2.90	2.56	2.69	2.33
There is not much room for my own ideas in math	3.02	3.11	3.00	3.04	3.09	3.00	2.64	2.72	2.55	2.60	2.70	2.52	2.59	2.30
I am challenged by math problems I can't understand immediately	3.44	3.42	3.36	3.29	3.31	3.33	3.34	3.27	3.25	3.07	3.26	3.01	3.27	3.19
The challenge of math problems does not appeal to me	2.67	2.93	3.00	3.05	2.95	3.14	3.05	2.86	2.85	2.89	3.10	2.96	3.18	3.11
When a math problem arises that I can't immediately solve, I stick with it until I have the solution	3.79	3.61	3.49	3.33	3.48	3.37	3.27	3.10	3.12	2.96	3.12	3.05	2.98	3.16
I do as little work in math as possible	3.96	4.21	3.72	3.96	3.73	3.86	3.70	3.71	3.63	3.56	3.47	3.43	3.10	3.56

[a]Higher score reflects more positive attitude about mathematics.

TABLE 4.
Student Means on the Useful/Useless Scale

ITEM MEANS[a]

ITEMS	Grade 6 M	Grade 6 F	Grade 7 M	Grade 7 F	Grade 8 M	Grade 8 F	Grade 9 M	Grade 9 F	Grade 10 M	Grade 10 F	Grade 11 M	Grade 11 F	Grade 12 M	Grade 12 F
It is important to know math in order to get a good job	4.18	3.98	4.00	3.75	3.97	3.76	4.04	3.69	3.67	3.53	3.56	3.33	3.22	3.23
Outside of science and engineering, there is little need for mathematics in most jobs	3.69	3.60	3.82	3.95	3.78	3.85	3.91	3.80	3.86	3.90	3.89	3.83	3.76	3.97
Mathematics is a worthwhile and necessary subject	4.23	4.27	4.22	4.04	4.22	4.16	4.24	3.89	4.04	3.94	4.14	3.89	3.82	3.86
Taking mathematics is a waste of time	4.32	4.26	4.22	4.26	4.23	4.33	4.19	4.19	4.04	4.12	4.06	4.07	4.06	4.14
I study math because I know how useful it is	3.85	3.73	3.86	3.63	3.73	3.62	3.74	3.34	3.65	3.35	3.59	3.32	3.41	3.33
Math is not useful for the problems of everyday life	3.58	3.57	3.72	3.80	3.60	3.77	3.67	3.66	3.59	3.69	3.69	3.65	3.71	3.64
Knowing mathematics will help me earn a living	4.32	4.12	4.31	3.96	4.28	4.09	4.30	3.82	4.08	3.74	3.98	3.74	3.79	3.57
Math will not be important to me in my life's work	4.12	3.97	3.96	3.86	4.00	3.88	4.02	3.62	3.84	3.62	3.88	3.48	3.69	3.59
I will use math in many ways as an adult	4.10	3.73	3.88	3.82	4.04	3.91	3.95	3.59	3.91	3.73	3.84	3.64	3.57	3.54
I expect to have little use for math when I get out of school	3.77	3.63	3.92	3.78	3.97	3.73	3.99	3.73	3.90	3.62	3.91	3.51	3.70	3.58

[a]Higher score reflects more positive attitude about mathematics.

TABLE 5.
Student Means on Support/No Support
from Others Scale

ITEM MEANS[a]

ITEMS	Grade 7		Grade 8		Grade 10		Grade 11	
	M	F	M	F	M	F	M	F
My parents expect me to do well in math	4.24	3.88	4.19	3.97	4.01	3.95	3.94	3.82
My parents do not think that I can do well in math	4.07	4.07	4.03	4.18	4.10	4.07	3.93	3.90
My friends support me in my study of math	2.73	2.72	2.79	2.91	3.02	2.88	2.84	2.97
My friends don't support me in my study of math	2.96	3.18	2.95	3.15	3.18	3.06	3.09	3.11
My teachers encourage me in my study of math	3.71	3.50	3.64	3.49	3.55	3.27	3.60	3.34
My teachers are discouraging about my ability in math	3.74	3.49	3.80	3.73	3.60	3.66	3.70	3.54
My friends believe that I am good in math	3.19	2.98	3.29	3.16	3.10	2.97	3.14	2.93
My friends don't think of me as a math whiz	2.46	2.18	2.61	2.54	2.36	2.20	2.50	2.34
My parents encourage me in my study of mathematics	4.07	4.05	3.95	3.90	4.01	3.77	3.88	3.74
My parents do not encourage me in my study of math	4.18	4.19	4.00	4.18	4.18	4.17	4.04	4.09
My teachers think I can do well in math	3.76	3.38	3.82	3.63	3.74	3.45	3.65	3.33
My teachers do not expect me to do well in mathematics	3.96	3.92	3.86	3.93	3.77	3.90	3.82	3.59

[a]Higher score reflects more positive attitude about mathematics.

TABLE 6.

Student Means on Math is Open to All/A Male Domain Scale

	ITEM MEANS[a]							
ITEMS	Grade 7		Grade 8		Grade 10		Grade 11	
	M	F	M	F	M	F	M	F
I would trust a woman just as much as I would trust a man to figure out important calculations	4.12	4.27	4.16	4.49	4.09	4.50	4.18	4.48
I would have more faith in the answer for a math problem solved by a man than a woman	3.65	4.23	3.83	4.35	3.75	4.38	3.84	4.38
Women are certainly logical enough to do well in mathematics	3.91	4.09	3.98	4.25	3.89	4.07	3.94	4.12
I would expect a woman mathematician to be a masculine type of person	3.62	3.79	3.68	4.10	3.86	4.30	3.98	4.32
Males are not naturally better than females in mathematics	3.19	3.71	3.28	3.56	3.29	3.48	3.31	3.64
Boys tend to be better at math than girls	3.55	4.20	3.69	4.31	3.54	4.13	3.67	4.19
Girls can do just as well as boys in mathematics	3.97	4.49	4.06	4.46	4.12	4.46	4.06	4.40
It's hard to believe a female could be a genius in mathematics	4.07	4.52	4.10	4.51	4.30	4.52	4.37	4.56
Females are as good as males at working problems about fractions	3.90	4.13	3.89	4.33	3.82	4.22	3.88	4.24
Girls who enjoy studying math are a bit peculiar	3.70	4.13	3.69	4.11	3.56	4.21	3.80	4.27
Studying mathematics is just as appropriate for women as for men	4.05	4.27	4.02	4.31	4.00	4.31	4.09	4.29
When a woman has to solve a math problem, it is appropriate to ask a man for help	3.50	4.05	3.53	4.10	3.61	4.24	3.72	4.18

[a]Higher score reflects more positive attitude about mathematics.

TABLE 7.
Correlations of Attitude Scales

FIRST-YEAR DATA: 514 STUDENTS

	Easy	Enjoyable	Creative	Useful
Like/Dislike	.58	.52	.58	.34
Easy/Difficult	—	.73	.68	.48
Enjoyable/Anxiety-Provoking		—	.65	.46
Creative/Dull			—	.46
Useful/Useless				—

SECOND-YEAR DATA: 514 STUDENTS

	Easy	Enjoyable	Creative	Useful	Support	Open	Distance
Like/Dislike	.64	.58	.64	.35	.42	.10	−.32
Easy/Difficult	—	.70	.66	.43	.56	.11	−.26
Enjoyable/Anxiety-Provoking		—	.67	.40	.45	.04	−.30
Creative/Dull			—	.48	.52	.13	−.42
Useful/Useless				—	.46	.25	−.28
Support/No Support from Others					—	.26	−.22
Math is Open to All/A Male Domain						—	−.09
Distance of Self/Mathematician[a]							—

THIRD-YEAR DATA: 514 STUDENTS

	Easy	Enjoyable	Creative	Useful	Support	Open	Distance
Like/Dislike	.64	.63	.65	.42	.47	.05	−.28
Easy/Difficult	—	.70	.67	.46	.59	.09	−.27
Enjoyable/Anxiety-Provoking		—	.70	.37	.46	−.03	−.33
Creative/Dull			—	.54	.52	.09	−.32
Useful/Useless				—	.51	.23	−.32
Support/No Support from Others					—	.27	−.21
Math is Open to All/A Male Domain						—	−.02
Distance of Self/Mathematician[a]							—

[a] A smaller distance is the positive end of this scale, so correlations of it with other scales should be negative.

TABLE 8.
Correlations of Subscales EASY, FUN, and FEAR

FIRST-YEAR DATA[2]	Easy	Enjoyable	Creative
EASY	—	.71	.64
FUN		—	.55
FEAR[b]			—
SECOND-YEAR DATA[a]	Easy	Enjoyable	Creative
EASY	—	.63	.65
FUN		—	.52
FEAR[b]			—
THIRD-YEAR DATA[a]	Easy	Enjoyable	Creative
EASY	—	.65	.70
FUN		—	.58
FEAR[b]			—

[a]N = 514 students

[b]This scale is calibrated so that a high score means little fear of mathematics, so correlations with other scales should be positive.

TABLE 9.
Correlations of Derived Measures for Each Year of Testing

FIRST-YEAR MEASURES[a]	Gender	SES	Ability	Feelings	Usefulness			
Gender[b]	—	.03	.06	.11	.10			
SES	.02	—	.45	.16	.07			
Ability	.02	.41	—	.18	.12			
Feelings	.24	.04	.17	—	.51			
Usefulness	.21	.07	.17	.47	—			
SECOND-YEAR MEASURES	**Gender**	**SES**	**Ability**	**Feelings**	**Usefulness**	**Support**	**Open**	**Distance**
Gender	—	.03	.06	.16	.08	.16	− .30	n.a.
SES	.02	—	.45	− .01	− .03	.03	.13	n.a.
Ability	.02	.41	—	.11	.10	.12	.35	n.a.
Feelings	.18	− .04	.08	—	.46	.47	.10	n.a.
Usefulness	.10	− .03	.10	.42	—	.38	.27	n.a.
Support from Teachers	.07	.04	.15	.44	.35	—	.20	n.a.
Math is Open to all	− .36	.16	.31	.06	.27	.25	—	n.a.
Distance of Self/ Mathematician[c]	− .03	.06	− .03	− .33	− .28	− .16	− .09	—
THIRD-YEAR MEASURES	**Gender**	**SES**	**Ability**	**Feelings**	**Usefulness**	**Support**	**Open**	**Distance**
Gender	—	.03	.06	.14	.00	.07	− .35	n.a.
SES	.02	—	.45	.02	.04	.13	.11	n.a.
Ability	.02	.41	—	.06	.16	.22	.26	n.a.
Feelings	.21	.06	.14	—	.33	.43	.05	n.a.
Usefulness	.12	− .04	.14	.48	—	.31	.32	n.a.
Support from Teachers	.09	.07	.20	.54	.39	—	.26	n.a.
Math is Open to all	− .41	.27	.37	− .03	.16	.19	—	n.a.
Distance of Self/ Mathematician[c]	− .05	.13	.00	− .28	− .32	− .05	.02	—

[a]Sixth-grade correlations appear in the upper right section of the table, ninth-grade in the lower left section.

[b]Male was coded as 1, female as 0, so positive correlations with gender imply males have higher scores.

[c]A lower distance between self and mathematician is "good," so negative correlations should be expected with other attitude measures.

TABLE 10.
Correlations of Measures Used in Regression Analyses

TWO-YEAR ANALYSES[a]	L.Feel	C.Feel	L.Use	C.Use	L.Teach	C.Teach	L.Male	C.Male	L.Self	C.Self	Ability	SES	Gender
Level of Feelings	—	.03	.41	-.01	.49	.05	.07	-.01	n.a.	n.a.	.10	.00	.16
Change in Feelings	.06	—	-.14	.31	-.03	.34	-.05	.11	n.a.	n.a.	-.05	.03	-.02
Level of Usefulness	.46	.06	—	.01	.41	.06	.31	-.01	n.a.	n.a.	.15	.01	.04
Change in Usefulness	.06	.43	.04	—	.05	.17	.06	.24	n.a.	n.a.	.06	.07	-.08
Level of Teacher Support	.49	.08	.41	.04	—	.13	.30	-.05	n.a.	n.a.	.21	.10	.14
Change in Teacher Support	.15	.54	.09	.30	.21	—	.12	.04	n.a.	n.a.	.11	.10	-.06
Level of Math is Male Domain	-.02	-.07	.20	-.08	.26	.05	—	.00	n.a.	n.a.	.33	.13	-.35
Change in Math is Male Domain	-.05	.18	-.02	.27	.08	-.10	.09	—	n.a.	n.a.	-.10	-.02	-.06
Level of Self/Mathematician[b]	-.34	-.10	-.38	-.07	-.12	.06	-.04	-.01	—	n.a.	n.a.	n.a.	n.a.
Change in Self/Mathematician[b]	.04	-.15	-.01	-.13	.15	.04	.08	.02	.15	—	n.a.	n.a.	n.a.
Ability	.12	.07	.13	.04	.22	.08	.37	.10	-.04	.01	—	.45	.06
SES	.01	.11	.01	.08	.07	.04	.24	.16	.11	.07	.41	—	.03
Gender	.22	.04	.12	.03	.10	.03	-.42	-.09	-.05	.03	.02	-.02	—

THREE-YEAR ANALYSES	Lev.Feel	Lin.Feel	C.Feel	Lev.Use	Lin.Use	C.Use	Ability	SES	Gender
Level of Feelings	—	.01	-.03	.48	-.09	.01	.14	.06	.16
Linear Feelings	.10	—	-.12	-.12	.33	.01	-.11	-.14	.02
Curved Feelings	-.03	-.04	—	-.07	-.12	.35	.02	.14	-.04
Level of Usefulness	.50	-.02	.07	—	-.04	-.03	.16	.04	.07
Linear Usefulness	.12	.33	.06	.17	—	-.06	.03	-.03	-.09
Curved Usefulness	.00	.02	.40	-.07	-.02	—	.05	.10	-.04
Ability	.15	-.02	.09	.16	-.01	.05	—	.45	.06
SES	.02	.03	.12	.03	-.02	.10	.41	—	.03
Gender	.25	-.02	.06	.17	-.06	.08	.02	.02	—

[a] Correlations for middle school students are given in the upper right of the table, for high school students in the lower left.

[b] These distances scores are the only measures for which a low score is more desirable. Negative correlations are therefore expected.

Index

Ability in mathematics
  course plan participation prediction
    with, 76–78, 86, 92, 93, 98
  self-perceptions of lack of, 9–10
Abrego, M. B., 12, 20
Achievement tests
  research project use of, 23–24, 36–37
  sex differences in mathematical ability
    on, 10
Adams, S., 12
Ahlgren, A., 18
Aiken, L. R., 10, 12, 13, 15, 16, 18, 20
Algebra, 50, 51
American Mathematical Society, 5, 107
Analysis of variance, 40–41
Anastasi, A., 10

Anderson, K. E., 18
Anttonen, R. G., 12
Anxiety
  questionnaire scales on, 31–33
  as reaction to mathematics, 13–15
  student essays on, 52–53, 54
Aptitude tests, 10
Arithmetic
  student like/dislike measure of, 50, 51
  usefulness of, in later life, 57
Armstrong, Jane A., 16, 20
Astin, H. A., 10
Attitudes toward mathematics, 38–70
  correlations to attitude scales, 155
  course plan participation prediction
    with, 78–81, 85, 92, 94

easy/difficult measure in, 43, 44, 46, 47, 48, 53
compared with attitudes toward English courses, 46
enjoyable/anxiety-provoking measures in, 43, 45, 46, 47, 48, 53
like/dislike measure in, 40–41, 44, 46, 47, 48
negative, 12–15
perceived usefulness of mathematics in, 15–17, 54–62
pleasure/displeasure measures in, 42–54
on questionnaire, 31–33
reasons for liking or disliking mathematics in, 49–51
suggestions for change in teaching and, 51, 52–53
topics especially liked or disliked and, 50, 51

Bachman, A. M., 13
Backman, M. E., 10
Bandura, A., 108
Bassham, H., 12
Beardslee, D. C., 18, 34
Bee, H., 18
Bem, S. L., 18
Berglund, G. W., 10, 15, 16
Biggs, J. B., 14, 20
Biology, student ranking of, 102, 103
Black children, 25
Blum, M. P., 12, 20
Boswell, S. L., 18
Boys. See Sex differences
Broverman, D., 18
Broverman, I., 18
Brown, J. D., 13
Brush, L. R., 14, 15, 18, 20
Bureau of Census, 25

Calculus, 5
Callahan, L. G., 10
Callahan, W. J., 12, 20
Campbell, N. J., 12
Card Rotation Test, 11
Career planning
future changes in career and, 62
job market and salaries in, 7
lack of mathematical preparation and restrictions in, 1–2
occupational flexibility and, 7–8
on questionnaire, 31
student choice of area in, 58, 59–60
student perception of level of education required for, 58–59, 60
usefulness of mathematics attitudes in, 15–17, 56–57

working mothers and, 8
Carey, G. L., 12, 15
Casserly, P. L., 18
Centra, J. A., 5
Chemistry, in student ranking, 102, 103
Classroom, and change, 99, 101–106
College degree, perceived need for, 57
College departments, entry into, 5
Cooper, D., 12
Course plans
abilities and, 76–78
attitudes in, 78–81
background variables in predicting, 75–76
introducing new, 100, 108–111
predicting participation in, 71–96
results from predicting, 83–95
sex differences in, 72–75
Creative/dull measures
as course plan predictor, 78
on questionnaire, 45, 46, 48
student means on, 151

Dawson, J. A., 12
Degnan, J. A., 12
Delon, F. G., 13
Denham, S. A., 16
Department of Labor, 7, 8, 26
Differential Aptitude Tests (DATs), 36–37, 77
Doctorates, sex differences in granting of, 5
Dreger, R. M., 12, 15, 20
Duckworth, D., 12
Dutton, W., 12, 15, 20

Easy/difficult measures
as course predictor, 78, 80
on questionnaire, 43, 44, 46, 47, 48, 53
student means on, 145
Edie, C. A., 20
Educational Testing Service (ETS), 3
Elementary schools, 15
Employment
mathematical skills needed in, 7
occupational flexibility in, 7–8
Engineering graduates, 7
English courses
attitude and anxiety scales on questionnaire on, 32–33
comparison between mathematics classes and, 104–105
requiring four years of, 111–112
student attitudes toward, 46, 54
student ranking of, 102, 103
Enjoyable/anxiety-provoking measures
as course plan predictor, 78, 80

on questionnaire, 43, 45, 46, 47, 48, 53
 student means on, 150
Entwistle, N. J., 12
Ernset, J., 18
Essays, 35
 influences from social milieu on, 67
 perceived usefulness of mathematics
  in, 56
 reasons for liking and disliking
  mathematics in, 48–52
 suggestions for change in teaching in,
  51, 52–53
 topics especially liked or disliked in,
  50, 51

Factor analysis, 78
Family, and career choice by girls, 68
Family background, on questionnaire, 30
Fedon, J. P., 12
Feelings. See Attitudes about mathematics
Fellows, M. M., 12, 13
Fennema, E., 10, 12, 13, 14, 15, 16, 18, 32,
 33
Flanagan, J. C., 10
Fox, L. H., 10, 16, 18

Garai, J. E., 10
Gender
 in predicting course plan participation,
  75, 85–86, 89, 92, 93
 See also Sex differences
Genetic factors, 3
Geometry
 innovative teaching of, 105–106
 student like/dislike measures of, 50, 51
Gestalt Completion Test, 11
Gilbert, C. D., 12
Girls
 career choices and family, importance
  for, 68
 perceived usefulness of mathematics in
  lives of, 16, 56–57
 value placed on family by, 68
 See also Sex differences
Glennon, V. J., 10
Gough, M. F., 10
Grade differences
 course plans in mathematics and, 73,
  75
 in course preferences, 102, 103
 mathematics as open to all or male
  domain and, 65
 perceived usefulness of mathematics
  and, 54–62
 pleasure/displeasure measures and, 48
Guidance counselors, 107–108

Haven, E. W., 16, 19
Hidden Patterns Test, 11
High school diploma, perceived need for,
 57, 60
High school students
 course plan participation by, 92–95
 requiring four years of mathematics
  for, 100–101, 111–112
 in research sample, 25
 sex differences in enrollment in
  mathematics courses by, 4, 5, 6
 support from parents and teachers
  perceived by, 65
 topics especially liked or disliked by,
  50, 51
Hilton, T. L., 10, 15, 16
History classes, in student ranking, 102,
 103
Hollingshead, A. B., 25
Holly, K. A., 12
Hudson, L., 18
Humanities graduates, 7
Hungerman, A. D., 13, 20
Hunkler, R., 13
Husen, T., 13, 15, 16, 32

Interviews
 influences from social milieu on, 67–68
 perceived usefulness of mathematics
  in, 56
 in research study, 23, 35–36
IQ scores
 course plan participation prediction
  with, 76–77
 in research study, 23–24, 36, 37

Jacklin, C. N., 10, 54
Jacobson, L., 19

Kaminski, D. M., 18
Keeling, B., 20
Koff, E., 18
Kreinberg, N., 5, 107

Language studies, in student ranking, 102,
 103
Lazarus, M., 20
Like/dislike measures
 as course plan predictor, 78, 80
 in research study, 40–41, 44, 46, 47, 48
Lyda, W. J., 13

Maccoby, E. E., 10, 54
McCallon, E. L., 13
McCarthy, J. L., 5
McNarry, L. R., 18
Massachusetts Institute of Technology, 107

Math/Science Network, 107
Mathematics courses
    comparison between English courses
        and, 104–105
    in student ranking, 102–103
Mathematics Percentile, in course plan
    participation prediction, 76, 77
Mathematics study
    exploration of avoidance in, 9–19
    perspective on world from, 8–9
    self-perception of lack of ability in,
        9–10
Michael, W. B., 13
Middle school students
    course plan participation by, 5, 85–91
    in research sample, 25
    support from parents and teachers
        perceived by, 65
Minority children, 25
Mokron, J. R., 18
Morse, E. C., 13
Mothers. See Working mothers
Mullis, I. V. S., 10
Murphy, K., 12
Murphy, M., 12
Muscio, R. D., 13

National Assessment of Educational
    Programs (NAEP), 3
Nicoletti, J., 20
Norton, D. A., 19
Number anxiety, 13

Occupational Outlook Handbook
    (Department of Labor), 7, 57
O'Dowd, D. D., 18, 34
O'Farrell, S., 18

Parents
    designing programs for, 100, 108
    support for mathematics from, 18–19,
        62–65
Parsons, J., 10
Perl, T., 107
Personality, and choice of mathematics, 18
Physics courses, in student ranking, 102,
    103
Poffenberger, T., 19
Project EQUALS, 107
Project WAM, 107
Project WITS, 107
Purl, M. C., 12

Quast, W. G., 13
Questionnaire, 22–23, 29–34
    categories on, 30–34
    influences from social milieu on, 62–67

pleasure/displeasure measures on,
    42–54
test of, 131–145

Redlich, R. C., 25
Research project, 21–37
    essays in, 35
    interviews in, 23, 35–36
    questionnaire used in, 22–23, 29–34,
        131–145
    students in, 24–28
    test scores in, 23–24, 36–37
Reys, R. E., 13
Richardson, F. C., 20, 32
Roberts, F., 13, 20
Rosenkrantz, P., 18
Rosenthal, R., 19

Salaries, and college field of study, 7
Sarason, I. G., 54
Scheinfeld, A., 10
Schoen, H. L., 12
Scholastic Aptitude Test (SAT), 3
School guidance counselors, 107–108
Schwandt, A. K., 5
Self-image, on questionnaire, 33–34
Sell, Lucy, 2–3, 5, 18, 20
Sepie, A. C., 20
Sex differences
    answering patterns on questionnaire
        and, 53–54
    area of desired career and, 58, 59–60
    course plans in mathematics and, 73
    in doctorates granted, 5
    genetic factors and, 3
    like/dislike measures and, 40–41
    mathematics as open to all or male
        domain and, 65
    measures of mathematical ability and,
        10–12
    negative attitudes toward mathematics
        and, 15
    perceived support from parents and
        teachers and, 62–65
    perceived usefulness of mathematics
        and, 55–56
    personality and choice of mathematics
        and, 18
    plans to study mathematics and, 3–5, 6
    pleasure/displeasure measures and, 43,
        48, 53–54
    stereotyping of mathematics and male
        domain and, 17–18
    student perception of level of
        education required for desired
        career, 58–59, 60
    student perception of ability and, 10

support and expectations from
  environment and, 18–19
  usefulness of mathematics attitudes
  and, 16, 17
Sheehan, T. J., 10
Sherman, J. A., 10, 12, 13, 14, 15, 16, 18,
  32, 33
Smith, F., 13
Smithells, J., 18
Social milieu influences, 62–68
  enrollment in mathematics courses
  and, 17–19
  self-concept and choice of
  mathematics and, 18, 64, 66–67
  stereotyping of mathematics as male
  domain and, 17–18, 63, 65–66
  support from parents and teachers for
  mathematics in, 18–19, 62–65
Social studies, in student ranking, 102, 103
Socioeconomic status (SES)
  course plan participation prediction
  with, 75–76, 86, 89, 92, 93
  research project and, 25–26
Spatial achievement scores
  course plan participation prediction
  with, 76, 77
  in research project, 23–24, 37
  sex differences in, 10–12
Spinelli, P. R., 20
Stein, A. H., 18
Stereotyping of mathematics as male
  domain, 17–18, 33–34
Stright, V. M., 13, 20
Suinn, R. M., 20, 32
Support/no support scale
  course plan participation prediction
  with, 78, 80
  student means on, 153

Teachers
  designing programs for, 100, 108

questions on questionnaire about, 23
  student career aspirations and, 60–62
  student self-perception of ability and,
  10
  support for mathematics performance
  from, 18–19, 62–65
Teaching, student suggestions for changes
  in, 51, 52–53
Tobias, S., 20
Todd, R. M., 13

University of California at Berkeley, 3, 5,
  107
Usefulness of mathematics attitudes
  career choice and, 15–17, 56–57
  course plan participation prediction
  with, 78, 80, 86, 88
  student means on, 152

Verbal achievement scores
  course plan participation prediction
  with, 76, 77
  in research project, 23–24, 36–37
Vogel, S., 18
Von Brock, R. C., 12

Walberg, H., 18
Wellesley College, 110
White-collar families, 76
Wilson, J. W., 10
Winkle, G. H., 54
Wise, L. L., 16
Wolfle, D., 5
Women and Mathematics (WAM), 107
Women in Technology and Science (WITS),
  107
Women's Bureau, 25
Working-class families, 26, 76
Working mothers
  career preparation and, 8
  in research sample, 26–27